Mathematical Investigations: An Introduction to Algebraic Thinking

Concepts and Processes for the College Student

Phil DeMarois, *William Rainey Harper College*

Mercedes McGowen, *William Rainey Harper College*

Darlene Whitkanack, *Indiana University*

Sponsoring Editor: Karin E. Wagner
Developmental Editor: Kathy Richmond
Project Editor: Lisa A. De Mol
Design Administrator: Jess Schaal
Cover Design: Lesiak/Crampton Design: Lucy Lesiak
Cover Photo: Letraset Phototone
Production Administrator: Randee Wire
Printer and Binder: R.R. Donnelley & Sons Company
Cover Printer: Phoenix Color Corporation

Mathematical Investigations: An Introduction to Algebraic Thinking, Preliminary Edition

Partial support for this work was provided by the National Science Foundation's Course and Curriculum Development Program through grant DUE #9354471. The opinions, findings, and conclusions or recommendations expressed in this publication are those of the authors and do not necessarily reflect the views of the National Science Foundation.

This book was produced using FrameMaker 4.0.2.

Library of Congress Cataloging-in-Publication Data
DeMarois, Phil
Mathematical investigations : an introduction to algebraic thinking : concepts and processes for the college student / Phil DeMarois, Mercedes McGowen, Darlene Whitkanack. -- Prelim. ed.
p. cm.
Includes index.
ISBN 0-06-501288-7
1. Algebra. I. McGowen, Mercedes A. II. Whitkanack, Darlene.
III. Title.
QA154.2.D446 1996
512.9--dc20 95-16283
CIP

95 96 97 98 9 8 7 6 5 4 3 2 1

Table of Contents

Table of Contents

Preface

The Course and its Audience

This book is the first in a series which will provide curriculum materials for teachers who want to make changes in the way they teach algebra. We wanted to teach an algebra course that focused on function and concepts rather than rote skills, but we found the materials available did not meet our needs for several reasons. Sometimes they were written for more mathematically mature students than were enrolled in college developmental algebra courses. Other materials did not fully integrate the tools that should become as automatic as pencil and paper to enhance the student's ability to explore mathematics. Most important, many of the materials seemed in conflict with mathematics education research on the teaching and learning of algebra. This book is not for the faint–hearted. Teachers who have piloted these materials report that their use changed the way they thought about mathematics and the way they teach. Knowing something about the background and philosophy of these materials may help those contemplating the use of non–traditional materials in their classroom.

The draft of the National Council of Teachers of Mathematics (NCTM) Standards gave direction to change the way mathematics was taught. With the inclusion of technology use as an essential feature, the emphasis on problem solving, communication, reasoning, and connections empowered students in the field of mathematics and gave hope to frustrated teachers. Perhaps there was a way to make mathematics accessible to all students who were disenfranchised because of their lack of previous long–term success. Documents from the National Research Council, the Mathematical Sciences Education Board (MSEB) and the Mathematical Association of America (MAA) reinforced the need for change. We began to develop a set of materials that reflect our interpretation and agreement with ideas coming from research and working conferences. The draft of the American Mathematical Association of Two–Year Colleges (AMATYC) Standards has validated our efforts.

The changes called for in these documents focus attention on the need for restructured curriculum materials coupled with changes in instructional philosophy and strategy. Change occurs when teachers have a plan to achieve that change and transition materials to help all students make the change confidently. The materials must address appropriate technology use, learning styles, classroom management, assessment, and core curriculum.

An important aspect is the need for relevant content. Developmental algebra courses which meet the needs of students and industry can no longer be based on algorithmic techniques. Material must be presented in context. The sequence of ideas must be carefully considered. What is valued should be assessed but not necessarily graded. More emphasis should be placed on algebraic reasoning. A long–term immersion in language and structure, acquired and used over time, becomes an essential element of studying algebra. This kind of change does not occur magically. Teachers must have support and sufficient background to make these changes. Carefully planned in-service and training to refocus and support teachers in acquiring a broader view of algebra is essential.

Approach

This text promotes a pedagogical approach based on a constructivist perspective of how mathematics is learned. A student learns mathematics by working in a social context to construct mathematical ideas, and to reflect on these constructions, all as a means of making sense out of problem situations. In our book each section of the chapter begins with an Investigation of a problem situation. After gathering data, students work collaboratively on small tasks based on the Investigation activities. The tasks help students reflect on mental constructions they have made as a result of working with the calculator and/or computer. Having students talk with the members of their group as they work on the task is another way of getting them to reflect on the problems and the solutions, whether discovered by themselves or presented by someone else in class (such as a member of another group or even the instructor). Students are expected to answer each Investigation in the text—*they write* portions of the book. A discussion in the text summarizes essential mathematical ideas. The instructor orchestrates intergroup and class discussions of the Investigation and the tasks. Explorations are included to reinforce the knowledge students are expected to have constructed during the first two steps of the cycle.

There is a significant difference between our approach and that of traditional materials: we avoid assigning exercises until *after* there is a reason to believe that students have constructed the relevant knowledge. When one is learning something new, there is a tendency for early interpretations to be inappropriate as students overgeneralize. Working with too many similar examples could cast these inadequacies in stone. With our approach, we hope to reinforce understandings and not *mis*understandings.

Distinguishing Features of the Text

S P C

College-level introductory algebra classes contain students with varying mathematical backgrounds and abilities. Introducing the S P C investigations in the beginning unit places students with weaker algebraic backgrounds and more knowledgeable students at the same starting point.

Because of the uniqueness of the problem, all students approach the problem on an equal basis. Prior expertise or lack of it does not affect a student's ability to generate paths using the rules of the S P C system or to hypothesize. Students form a sense of community centered around their mutual discoveries that mathematics is present even in seemingly non-mathematical situations as they discuss and justify their ideas and observe how other students reason. They discover that they can think and do mathematics.

Teachers who have piloted the unit containing the S P C system report that this beginning unit has a positive impact on their students' general knowledge of what mathematics (or algebra) is and is not. Many students are convinced and satisfied that algebra is a set of rules for manipulating symbols and enter class with the expectation of following a textbook's algorithmic approach. While investigating the S P C system, they experience the creative side of doing mathematics. Students discover that "just because a rule exists doesn't mean you have to use it," "the converse of a rule is not necessarily a rule," and "mathematics is just a matter of being consistent."

Using the S P C puzzle in the beginning of the book is a clear statement to students and teachers alike: THIS IS NOT THE USUAL INTRODUCTORY ALGEBRA BOOK. This text treats adult algebra students as adult learners by showing respect for their life experiences and by placing appropriate demands on their cognitive level of development. The *process* of learning mathematics as well as the *nature of mathematical thinking* is introduced in Chapter One with the S P C system. Teachers who believe that the S P C investigations are too abstract and difficult do their students a disservice. They under-estimate their students' capabilities; in fact, this problem has been used successfully with middle grade and high school students as well as with students in developmental mathematics courses at the college level. Students not only successfully work through the S P C investigations but gain confidence in their own abilities to "do mathematics" that continues throughout the course. The limited view of algebra as "solving equations" changes as a result of experiences with S P C. "Algebra" implies a mathematical structure in which the manipulation of symbols is only one feature. In an effort to develop an appreciation of the underlying structure of algebra, various ways of thinking about algebra are explored, with the goal of having students exit the course understanding that algebra is much more than" using variables to solve equations". Students are introduced to the idea of structure through the study of a logical system, the S P C system. As new number systems are introduced, their structure is emphasized and compared to that of the S P C system. The integration of mathematical topics from algebra and number theory help students see connections, develop understanding and insights not previously experienced.

Multiple Representations

This introductory algebra book is based on multiple representations of the function concept. There are interactive investigations on calculator and computer designed to help students make mathematical discoveries, to support small group work both in and out of class, and to reduce the teacher's role as lecturer.

Discovery-based Learning

There are no illustrated examples followed by exercises that repeat the examples. The Investigations tend to ask more questions than the subsequent Discussion answers. Students are expected to create their own glossary of terms and definitions. Perhaps the most innovative feature of the text is that, very often, students are asked to explore mathematical ideas on the calculator and/or computer before these ideas are discussed in class. Although this is a departure from standard practice, both faculty and students who have used the materials feel that this is an important component of our approach to helping students learn mathematical concepts and develop confidence in their ability to direct their own learning. Not only does such a style give students an opportunity to discover mathematical ideas on their own, but even if they don't succeed in making the discovery, the activities form an experiential base that helps them understand an explanation presented in class by another student or by the instructor.

Concept Maps
Reflections

Two special features conclude each section. A Concept Map centered on some key idea in the section provides students with a chance to brainstorm about the idea and draw connections between various aspects of the concept. The second concluding feature is a Reflection which asks students to write a short narrative discussing important concepts within a section and between sections.

Another distinctive feature is that the materials include space to write in observations and work interspersed with the Investigations and Explorations in the text. Alternative forms of assessment include student journals, concept maps, reflections, and class notes. Students develop organizational skills and an ownership in their text as they actually contribute to the writing of the Discussion and Investigation sections during the semester.

Technology Focus

Students use graphing calculators regularly. The graphing calculator is a function machine that has the capability of displaying both input and output on the screen simultaneously so that students can obtain immediate feedback, discover patterns, and identify their previously–learned misconceptions. Students are expected to have access to a computer algebra system during the course. Investigations done on a computer algebra system allow students to discover symbolic patterns and formulate rules. The use of technology provides opportunities for students to understand and interpret data in terms of the use of tools and techniques which are appropriate to the job. The emphasis is on the development of decision making and problem solving. Students are encouraged to make good choices about the techniques which make the most sense in a particular situation and to check answers for reasonableness and accuracy. We currently use a graphing calculator with a dynamic table feature and a user–friendly software package with symbolic manipulation capability but are not dependent on a specific calculator or on specific computer software.

Skill Development

A key issue in algebra is the development of appropriate skills for later courses. However, we believe that problem solving skills which encourage independent learning are more important than rote manipulation skills taught out of context. We cover most skills traditionally included in an algebra course plus advanced skills because students confront interesting problems that require these skills. The focus is on interesting problems that require certain skills in the process of investigating the problem. We prefer not to cover skills out of context. We want to emphasize connections among mathematical ideas and skills. This requires a major change in sequencing. If student learning is connected, the chance for retrieval of the knowledge in other problem situations is much greater. Based on constructivist principles, students will develop reasoning from a numerical standpoint by looking for patterns. By generalizing these patterns, they will create mathematics that is their own—this holds true for both concepts and skills.

Instructor Support

A support system must be in place for those who choose to implement materials that are radically different. We have been awarded an NSF grant to aid us in developing workshops and an on–going network for faculty who use this text.

Acknowledgments

There are many people to whom we need to say “Thank you”. We would like to acknowledge specifically Frank Demana and Bert Waits for their pioneering efforts in promoting the widespread acceptance and use of technology to help students explore and visualize mathematics and who served as models for us.

Since the basis for much of the development of this text is grounded in the mathematics education research literature, we would also like to acknowledge the contributions of Jim Kaput and the members of the Algebra Working Group of the National Center for Research in Mathematical Sciences Education. Other researchers who have had a major impact on our work include Robert B. Davis, Ed Dubinsky, Jim Fey, Eddie Gray, Carolyn

Kieran, Keith Schwingendorf, David Tall, Zalman Usiskin, and Sigrid Wagner. To this dedicated, articulate group of individuals, and to all those whose research guided our efforts and shaped our vision about what algebra should be, we say "Thank you".

We also need to acknowledge the major contribution of all those who have field-tested draft versions of this text, along with their students. The insightful recommendations for improvement, editorial comments and energetic efforts of these pioneer field-testers and students have been invaluable in making the text more useful and student friendly.

Class Testers

Margaret R. Crider, *Tomball College*

Colette Currie, *National-Louis University*

Irene Duranczyk, *Eastern Michigan University*

Ernie East, *Northwestern Michigan College*

Deborah Faust, *William Rainey Harper College*

Edward Gallo, *Indiana University East*

Gail Johnson, *Siena Heights College*

Kathy Ketchie, *Rowan-Cabarrus Community College*

Roxann King, *Prince George's Community College*

Dan Loprieno, *William Rainey Harper College*

Debra Pharo, *Northwestern Michigan College*

Nancy Rice, *William Rainey Harper College*

Jon W. Scott, *Montgomery College*

Pat Stone, *Tomball College*

Lana Taylor, *Siena Heights College*

Jo Warner, *Eastern Michigan University*

Tom Williams, *Rowan-Cabarrus Community College*

William Worpenberg, *Indiana University East*

A remarkable group of colleagues, Lana Taylor, Gail Johnson, Nancy Rice and Colette Currie, deserve a special mention. They were willing to test a very early draft of these materials in their classrooms without first seeing a completed project, receiving a chapter at a time during that first semester back in Spring, 1992. There are no words that adequately acknowledge the gift of trust and support they gave to us in the initial stages of this writing project.

We greatly appreciate the detailed feedback and commentaries provided us by all the reviewers of the various drafts of this text. The reviews addressed specific points of concern, validated our vision of the direction we chose, and helped us clarify content issues.

Reviewers

Richard A. Butterworth, *Massasoit Community College*

Frances Campbell-LaVoie, *University of San Francisco*

Marilyn P. Carlson, *Kansas University*

Stephen A. Chiappari, *Avila College*

Linda Crabtree, *Longview Community College*

Lillie Crowley, *Lexington Community College*

Allen Davis, *Eastern Illinois University*

Maryann B. Faller, *Adirondack Community College*

Kathryn Gillespie, *Heartland Community College*

Judith A. Godwin, *Collin County Community College*

Elaine M. Hale, *Georgia State University*

William L. Hoard, *Front Range Community College*

Amelia S. Kinard, *Columbia College*

Jillian Knowles, *Merrimack College*

Pam Littleton, *Tarleton State University*

John W. McConnell, *Glenbrook South High School*

Robert A. Powers, *Front Range Community College*

Bobby M. Righi, *Seattle Central Community College*

Sylvester Roebuck Jr, *Olive-Harvey College*

Pamela Roland, *Middlesex Community Technical College*

Jane Sieberth, *Franklin University*

Courtney L. Small, *Robert Morris College*

Gwen H. Terwilliger, *University of Toledo*

Lynn E. Trimpe, *Linn-Benton Community College*

Karl M. Zilm, *Lewis & Clark Community College*

Finally, we also wish to acknowledge our editors, Anne Kelly and Karin Wagner, together with our publisher, for their support during the development of this text.

We invite you to join the on-going discussion among colleagues who are using these materials. We encourage you to share your experiences, both sucessful and not so successful, with others who are attempting to change, not only the content but their instruftional practices as well. Our email addresses are listed below. Contact us if you woul like to become part of the network.

Sincerely,

Phil DeMarois
pdemaroi@harper.cc.il.us

Mercedes McGowen
mmcgowen@harper.cc.il.us

Darlene Whitkanack
darlene155@aol.com

Chapter 1

What is Mathematics?

Section 1.1 Learning Mathematics

Purpose

- Investigate the process of thinking about the process of doing mathematics (***meta-cognition***)
- Explore a problem in several ways.

Investigation

All twenty parking spaces in my favorite parking lot are filled. Some are occupied by motorcycles and others by cars. Some people count to ten when they get angry, but that wasn't nearly far enough. I counted wheels–sixty–six to be exact. How many cars and how many motorcycles had invaded my territory?

You can do this problem. It is not important which technique or strategy you use. Follow your intuition and find some way to determine a solution. As you work alone on this problem, answer the following investigations **in** the text.

1. Have you ever worked a problem like this before?

2. Describe *in words* what you did to investigate the problem.

3. Are you sure your solution is correct? Why or why not? How do you normally determine if you have a correct solution?

4. Are you satisfied with *how* you investigated the problem? What information (assumption) do you need to know that is not given in order to solve the problem?

5. Now work with a partner. If you were stuck, don't just copy the solution of your partner. Ask questions and see if you can use the hints to solve the problem together. If you both have solutions, see if your answers are the same. Did you both work the problem using the same strategy?

Discussion

Questions like the parking lot problem are very interesting to mathematicians because there are several very different strategies which lead to a solution.

The following strategies might have been used on the given problem.

- ***Guess and test***: guess a solution and test to see if the answer matches all the conditions.
- ***Draw a picture***: can you visualize the parking lot? You don't have to draw the vehicles, just the wheels.
- ***Use logic and arithmetic computation***: identify the numbers in the problem. Use only numbers given or implied to find the solution.
- ***Make an organized list***: use a table to display information about the problem in an organized way.
- ***Write an equation***: use either the number of cars or the number of motorcycles as a variable to write an equation that represents the problem.
- ***Write a system of equations***: Use the number of cars as one variable and the number of motorcycles as a second variable. Write two equations that represent the relationships between the number of cars and the number of motorcycles.

Investigation

6. Working in groups, try to solve the problem using each of these strategies.

7. Which of the strategies seem more mathematical to you? List your criteria for mathematical.

8. Which strategy seems "best"? List your criteria for "best".

9. Which of the strategies would you like to learn more about?

Discussion

In reality there are many criteria for determining a good solution. Some of the criteria you have listed might include:

- Efficiency—can I solve the problem with a minimal amount of effort? "No pain–no gain" applies more to physical exercise. In this problem a lucky guess might have been very efficient. For some it might have led to a lot of wasted effort.

- Transferability—can the process be generalized to solve a wide variety of problems like this, even when the answers are fractions or worse?

- Comprehensibility—can I understand what I am doing and why? Techniques that are not understood will be forgotten or remembered inaccurately so they result in incorrect answers.

It is important to remember that you are never finished with a problem until you know *more* than just the steps to find a solution. You should be able to explain how the steps in the problem relate to the situation and why you want to use those steps.

Let's look at some methods of solution and variations on these methods.

- *Guess and test.*

  If you guessed ten cars and ten motorcycles (hopefully you used the fact that the total vehicles added up to twenty) and tested $(10 \times 4) + (10 \times 2) = 40 + 20 = 60$ wheels. This is too low (we need 66 wheels) so there must be more cars.

  If there are twelve cars and eight motorcycles we get $(12 \times 4) + (8 \times 2) = 48 + 16 = 64$ wheels.

  If there are thirteen cars and seven motorcycles we get $(13 \times 4) + (7 \times 2) = 52 + 14 = 66$ wheels. This is the correct solution.

  If you use a calculator, you are less restricted by what numbers you choose since the calculator makes the computations easy.

- *Draw a picture.*

  I'm not much of an artist but it doesn't matter. I only need to show parking spaces and wheels. We use the fact that there are twenty vehicles. We could have focussed initially on the number of wheels.

  First draw twenty spaces.

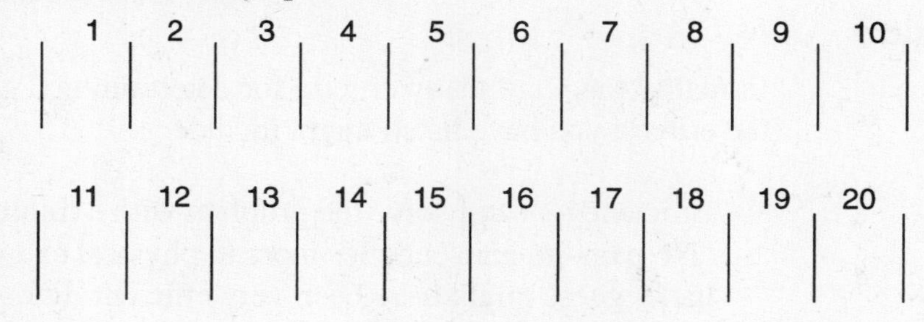

  Put two wheels in each space.

1 2 3 4 5 6 7 8 9 10
11 12 13 14 15 16 17 18 19 20

A variation would be to first put four wheels in every space for a total of eighty wheels.

Put two more in each as long as you can—until you reach 66 wheels.

1	2	3	4	5	6	7	8	9	10
OO OO	OO OO	OO OO	OO OO	OO OO	OO OO	OO OO	OO OO	OO OO	OO OO

11	12	13	14	15	16	17	18	19	20
OO OO	OO OO	OO OO	OO	OO	OO	OO	OO	OO	OO

The last diagram displays an answer of thirteen cars and seven motorcycles.

Notice the reasoning used (logic)—since every space was occupied, there were at least two wheels in each spot. This uses up forty wheels. Then adding two more to each space is continued until sixty–six is reached.

If you initially drew eighty wheels, then you subtract pairs of wheels in each space until you reach sixty–six wheels.

If you use twenty spaces, then you need to check for a total of sixty–six wheels. If you use sixty–six wheels, then you need to check for twenty vehicles.

Do a reality check.

Problem Solving Toolkit

- *Use logic and arithmetic computation.* A computational approach uses only the numbers given in the problem.

  20 spaces times two wheels each equals forty wheels.

  66 wheels total subtract 40 wheels used so far equals 26 extra wheels.

  $26 \div 2 = 13$ cars and $20 - 13 = 7$ motorcycles.

- *Make an organized list using a table.*

An organized list puts the sequence of guesses in an orderly fashion.

A table has explicitly labeled columns as shown in Table 1 .

Number of cars	Number of car wheels	Number of motor-cycles	Number of motor–cycle wheels	Number of vehi-cles	Number of wheels
20	80	0	0	20	80
19	76	1	2	20	78
18	72	2	4	20	76
17	68	3	6	20	74
16	64	4	8	20	72
15	60	5	10	20	70
14	56	6	12	20	68
13	**52**	**7**	**14**	**20**	**66**
0	0	20	40	20	40

Table 1: Cars and motorcycles

Notice that often an inspection of the list reveals a pattern or many patterns which can be used to make a table that does not include every possible entry. If there are multiple solutions, an organized table is more likely to reveal them than the guess and test method.

A variation on Table 1 would begin with zero cars and increase the number of cars by one as we proceed down the table. Note that Table 2 assumes that the number of motorcycles depends on the number of cars. Another variation would be to let “Number of motorcycles” be the first column. Then the number of cars would depend on the number of motorcycles.

- *Write an equation.*

To write an equation, we need a variable. What we will do is the process someone who knows algebra might follow. It is not necessary that you comprehend the process completely at this point. It is included because some students probably remember this process and tried to use it.

In order to write an equation, it helps to recognize patterns within the problem. The organized table is an excellent choice for pattern recognition. Table 2 is an alternative version of Table 1(page 6) in which the process for the computations is displayed instead of the final answer. Patterns are more easily recognized if the processes are written out.

Number of cars	Car wheels	Motor-cycles	M-cycle wheels	Total Vehicles	Number of wheels
20	4 (20)	20 – 20	2 (0)	20 + (20 – 20)	4 (20) + 2 (20 – 20)
19	4 (19)	20 – 19	2 (1)	19 + (20 – 19)	4 (19) + 2 (20 – 19)
18	4 (18)	20 – 18	2 (2)	18 + (20 – 18)	4 (18) + 2 (20 – 18)
17	4 (17)	20 – 17	2 (3)	17 + (20 – 17)	4 (17) + 2 (20 – 17)
16	4 (16)	20 – 16	2 (4)	16 + (20 – 16)	4 (16) + 2 (20 – 16)
15	4 (15)	20 – 15	2 (5)	15 + (20 – 15)	4 (15) + 2 (20 – 15)
14	4 (14)	20 – 14	2 (6)	14 + (20 – 14)	4 (14) + 2 (20 – 14)
13	**4 (13)**	**20 – 13**	**2 (7)**	**13 + (20 – 13)**	**4 (13) + 2 (20 – 13)**
0	4 (0)	20 – 0	2 (20)	0 + (20 – 0)	4 (0) + 2 (20 – 0)
c	$4(c)$	$20 - c$	$2(20 - c)$	$c + (20 - c)$	$4(c) + 2(20 - c)$

Table 2: Cars and motorcycles process for one variable

Let the letter c represent the number of cars. Since there are twenty vehicles, if there were ten cars there would be ten motorcycles. If there were twelve cars there would be eight motorcycles. If there were two cars there would be $20 - 2$ or 18 motorcycles. Thus if c is the number of cars then $20 - c$ is the number of motorcycles.

The equation is written using the number of wheels.

$$4(c) + 2(20 - c) = 66$$

Solving the equation leads to a solution of 13. So there are 13 cars and 7 motorcycles.

- *Write a system of equations.*

Sometimes it is easier to have a variable for each unknown quantity. We try to write as many equations as variables. Sometimes this isn't possible. Table 3 displays the processes used to complete Table 1(page 6), but unlike Table 2 (page 7), it assumes the use of two variables.

Cars	Car wheels	Motor-cycles	M-cycle wheels	Total Vehicles	Number of wheels
20	4 (20)	0	2 (0)	20 + 0	4 (20) + 2 (0)
19	4 (19)	1	2 (1)	19 + 1	4 (19) + 2 (1)
18	4 (18)	2	2 (2)	18 + 2	4 (18) + 2 (2)
17	4 (17)	3	2 (3)	17 + 3	4 (17) + 2 (3)
16	4 (16)	3	2 (4)	16 + 4	4 (16) + 2 (4)
15	4 (15)	5	2 (5)	15 + 5	4 (15) + 2 (5)
14	4 (14)	6	2 (6)	14 + 6	4 (14) + 2 (6)
13	**4 (13)**	**7**	**2 (7)**	**13 + 7**	**4 (13) + 2 (7)**
...	...	...	...	...	...
0	4 (0)	20	2 (20)	0 + 20	4 (0) + 2 (20)
c	$4(c)$	m	$2(m)$	$c+m$	$4(c)+2(m)$

Table 3: Cars and motorcycles process for two variables

For example, let c represent the number of cars and m represent the number of motorcycles. Column 5 in Table 3 provides a pattern for the first equation. column 6 of Table 3 provides a pattern for the second equation. The two equations are

$$\begin{aligned} c + m &= 20 \\ 4c + 2m &= 66. \end{aligned}$$

There are several ways these equations can be solved including substitution, elimination, graphing, using matrices, and using determinants. The wide variety of solution techniques includes most of the mathematics we will explore in this text. Furthermore it suggests that "algebra" may be only one tool of many in the mathematics toolkit. What is important is not just to learn isolated skills in solving equations, but that understanding the problem situation can lead to many intuitive and formal methods of solution.

Why then do we need "algebra"? What is the most important role that algebra plays in mathematics? As we have seen, it is not the solution to equations, since this can often be done by numerical means.

First, algebra provides a language and structure for mathematics. This notion of structure will be explored in the S P C problem in the next section and will be revisited throughout the book as you create a structure that encompasses whole numbers, integers, rational numbers, and real numbers along with the operations and properties of these number systems.

Second, algebra provides the tools to generalize problem solving. By using variables, a specific property like $2 + 3 = 3 + 2$ can be generalized to $x + y = y + x$ for any real number pair.

Furthermore, general equations provide techniques for solutions to a whole class of problems which can be done by hand or by technology.

Explorations

1. List the patterns you see in Table 1 (page 6).

2. Explain each part of the equation $4(c) + 2(20 - c) = 66$? Why are we multiplying by four and by two.

3. How would the equation in Exploration 2 change if we used the number of motorcycles as the variable?

4. As you attempted to solve the problem, which did you think about first: wheels or vehicles? How did this affect your approach to the problem?

5. Why do you think that one row of Table 1 (p. 6) is in bold face?

6. Describe the meaning of each equation in the system of equations

$$\begin{aligned} c + m &= 20 \\ 4c + 2m &= 66. \end{aligned}$$

7. During the first half of a basketball game, the Bulls scored 58 points. Surprisingly, these points came from two and three–point field goals—they made no free throws. The statistics for the first half showed that the Bulls shot 53% from the field having made 27 out of 51 shots. Use as many different problem solving strategies as you can to determine the number of two–point shots the Bulls made and the number of three–point shots the Bulls made.

Concept Map

Construct a concept map centered on the phrase **problem solving**.

Reflection

Summarize the process you used to investigate the parking lot problem. How did you get started? Describe all subsequent tasks in your investigation, using complete sentences.

Section 1.2 Thinking Mathematically

Purpose

- Discover the mathematics in a problem.
- Reflect on what was done to “solve” the problem.
- Make and test conjectures about the problem.
- Acquire understanding about the importance of notation.

Investigation

Problem situation: The **S**tudent **P**roduction **C**orporation makes television sets. Ernie trains new employees and schedules the tasks to be completed on the assembly line in the production of the TV’s. Each production must begin by pushing the **start–up button**. The rest of the production consists of **processes** (like connecting the wires in the control panel) and quality–control **checks** (such as turning on the set to see if it works or if faces are a sick green). To describe a sequence of production tasks, use the letter **S** to represent **start–up**, the letter **P** to represent **process**, and the letter **C** to represent a quality–control **check**. Thus the sequence of letters **S P** represents a start–up followed by a process. The sequence of letters which describes the steps in the production is referred to as the **production path**.

Ernie was given four rules which he could use to create or alter production paths. Let’s try each rule individually first!

1. **Rule 1**: You may create a new production path by repeating the sequence of processes and checks that follow the start–up. (This is useful when training a new employee like Bert. By repeating the process Ernie shows him, Bert can learn how to operate the equipment.)

   Example: S **P C C** —Rule 1→ S **P C C P C C**

   Apply Rule 1, if possible, once to each of the following paths by doubling the sequence which follows S.

   a. S P P

   b. S P P P P

   c. S P C P

2. **Rule 2**: If the production path ends with a process P, you may add a check, C, at the end. Rule 2 does not apply to production paths that already end in a quality–control check C since two checks in a row are unnecessary.

   Example: S P C C **P** $\xrightarrow{\text{Rule 2}}$ S P C C **P C**

   Apply just Rule 2, if possible, once to each of the following production paths.

   a. S P P

   b. S P P P P

   c. S P C P

   d. S P C

3. **Rule 3**: If a production path contains three consecutive processes, you may replace the three processes with a quality–control check. (This is necessary when planning Bert's work. He would like to just keep doubling the processes but his work is still sloppy so he needs a planned check on the quality of his work.)

   Example: S P **P P P** P C $\xrightarrow{\text{Rule 3}}$ S P **C** P C

   Given the combination of letters S P P P P P P P P, use Rule 3 once to replace any three consecutive P's with a C. List all the production paths that can be derived from S P P P P P P P P using only Rule 3.

4. **Rule 4**: If a production path contains two consecutive quality–control checks, you may remove the two quality–control checks to make a new production path. (Sometimes, the demand for product requires that the path be streamlined. It involves the risk that some defective product may be shipped, but with a deadline, eliminating quality–control checks is often a risk managers are willing to take.)

   Example: S P C C P —Rule 4→ S P P

   Apply Rule 4, if possible, once to each of the following production paths by removing two consecutive C's.

   a. S P C C

   b. S P C C P P P P C

   c. S P P C P P

Discussion

Check the results of your investigations. When you applied Rule 1 in Investigation 1 you repeated the entire production path following the start–up.

a. S P P —Rule 1→ S P P **P P**

b. S P P P P —Rule 1→ S P P P P **P P P P**

c. S P C P —Rule 1→ S P C P **P C P**

Applying Rule 2 requires us to do two things. First, does the production path end with a process? If it does, we can apply Rule 2. If it does not, Rule 2 cannot be applied.

In Investigation 2, the new production paths are:

a. S P P —Rule 2→ S P P **C**

b. S P P P P —Rule 2→ S P P P P **C**

c. S P C P —Rule 2→ S P C P **C**

d. The production path S P C does not end in a process so Rule 2 cannot be applied.

Applying Rule 3 once to S P P P P P P P P presents some choices. Since there are eight processes, we must decide which three consecutive processes will be replaced with a quality–control check. We obtain any of the following paths. A complete answer to Investigation 3 must include all of these.

S C P P P P P   S P C P P P P   S P P C P P P

S P P P C P P   S P P P P C P   S P P P P P C

A complete answer must include all possibilities.

Problem Solving Toolkit

Before applying Rule 4, we must look to see if the production path has two consecutive quality–control checks. If not, Rule 4 cannot be applied. If the valid path contains two consecutive quality–control checks, apply Rule 4 by removing the two consecutive checks. In Investigation 4, the paths follow.

a. S P **C C** $\xrightarrow{\text{Rule 4}}$ S P

b. S P **C C** P P P P C $\xrightarrow{\text{Rule 4}}$ S P P P P P C

c. S P P C P P does not contain two consecutive quality–control checks so Rule 4 cannot be applied.

In each of the preceding investigations, we started with a given production path. What if we are not given a starting path? We must have an initial production path. The chief executive officer (CEO) of the **S**tudent **P**roduction **C**orporation authorized a starting production path of a start–up followed by a process (**S P**). This given path **S P** is called the **initial path**. A production path that can be obtained by applying the rules to the initial path **S P** is called a **valid path**.

The process of applying the rules to the initial **S P** to produce new valid paths is called a ***derivation***.

There are two ways to derive a valid path.

- Begin with the initial path S P and apply the rules.
- Begin with a valid path you have previously derived from S P and apply the rules.

Figure 1 displays a sample derivation.

$$\text{S P} \xrightarrow[\text{Rule 1}]{} \text{S P P} \xrightarrow[\text{Rule 2}]{} \text{S P P C}$$

Figure 1

This derivation began with S P and created two new valid paths S P P and S P P C. Note that each application of a rule produces a new valid path. Below each arrow, we write the rule used to move from an old valid path to a new valid path.

In the process of introducing production paths and valid paths, we have used a notation that recorded:

- where we started (the initial path S P or a previously–derived valid path)
- the rule we applied
- the result we obtained (new valid paths)

This notation is introduced along with the rules of the Student Production Corporation (**S P C** for short) so that everyone in class uses the same notation and so that we can read each other's work. It is easy to understand what someone else has done and they can follow what you have done as well. Notation is an important part of studying mathematics. Analyze new notation by addressing the following standards.

Standards for mathematical notation

- The notation makes the process or concept easier to understand.
- The notation makes it easier to discuss the process and/or results with others.
- The notation is used consistently by everyone.
- The notation is efficient.

The notation introduced to derive valid paths for the S P C satisfies these standards.

Now that you are familiar with the rules, you are ready to explore valid production paths by using the rules in various combinations. To expertly manage production at the S P C, we must be able to derive valid paths. Furthermore, it has become a game of the employees to see how many different production paths they can use. That is the goal of the next investigation.

Investigation

5. Indicate the rule used under each arrow in the following derivation, using appropriate notation.

   S P $\longrightarrow$ S P P $\longrightarrow$ S P P P P $\longrightarrow$ S P P P P C

   $\longrightarrow$ S C P C $\longrightarrow$ S C P C C P C $\longrightarrow$ S C P P C

6. List each of the valid paths in Investigation 5.

7. Show that the production paths listed below are valid paths. Beginning with S P, write the derivation for each path.

   a. S P C

   b. S P P C

   c. S C P

   d. S P P C P P C

Discussion

The rules used in Investigation 5 appear in Figure 2.

S P $\xrightarrow{\text{Rule 1}}$ S P P $\xrightarrow{\text{Rule 1}}$ S P P P P $\xrightarrow{\text{Rule 2}}$ S P P P P C

$\xrightarrow{\text{Rule 3}}$ S C P C $\xrightarrow{\text{Rule 1}}$ S C P C C P C $\xrightarrow{\text{Rule 4}}$ S C P P C

Figure 2

The valid paths that result are

S P P	S P P P P	S P P P P C
S C P C	S C P C C P C	S C P P C.

Some possible derivations for the paths in Investigation 7 follow.

a. S P $\xrightarrow{\text{Rule 2}}$ S P **C**

b. S P $\xrightarrow{\text{Rule 1}}$ S P **P** $\xrightarrow{\text{Rule 2}}$ S P P **C**

c. S P $\xrightarrow{\text{Rule 1}}$ S P **P** $\xrightarrow{\text{Rule 1}}$ S P P **P P** $\xrightarrow{\text{Rule 3}}$ S **C** P

d. S P $\xrightarrow{\text{Rule 1}}$ S P **P** $\xrightarrow{\text{Rule 2}}$ S P P **C** $\xrightarrow{\text{Rule 1}}$ S P P C **P P C**

Notice that you have many choices of which rule to apply next. Different choices result in different derivations and paths. Perhaps your derivations were different. Did you try working backwards, asking yourself what rule was used on a preceding path to generate S P P C P P C?

Work backwards from the known answer.

Problem Solving Toolkit

Since P P C is repeated twice, the preceding path must have contained P P C once. Starting from S P P C P P C, work backwards to the previous path S P P C.

? $\longrightarrow$ S P P C $\longrightarrow$ S P P C P P C

Rule 1 was applied to S P P C to obtain S P P C P P C.

Now work backwards from S P P C. One thing to notice is the C. This could have come from either Rule 2 or Rule 3. Since the C is at the end of the path, it most likely came from applying Rule 2. Thus the previous path was S P P.

S P P $\longrightarrow$ S P P C $\longrightarrow$ S P P C P P C

Working backwards from S P P to S P requires the recognition that P is repeated twice. By Rule 1, we note that the previous path is S P. Fill in the rules and we are done with the derivation.

S P $\xrightarrow{\text{Rule 1}}$ S P P $\xrightarrow{\text{Rule 2}}$ S P P C $\xrightarrow{\text{Rule 1}}$ S P P C P P C

The elements S, P, and C, the initial path S P, the four rules, and the resulting valid paths are collectively called the ***S P C system***[1]. Now that you have generated valid paths, it is important to look back at what you have done and how you did it. This is a habit that applies to all your assignments, not just the S P C system. When you answered the investigations, you worked both *within* the S P C system and *outside* the S P C system. You work within the system when you derive valid paths. When you look back at what you have done and search for patterns, you are working outside the system.

Let's return to our new employee, Bert. It never occurred to his trainer, Ernie, that Bert would have a problem with the start-up procedure. How could someone turn the conveyor belt on backwards anyway?!! When the chaos (TV sets falling off the assembly line could be called chaos) attracted the attention of the CEO (remember, he made up the rules), he told Ernie that Bert was so inept he should have checked the system immediately after the start-up and stopped--that is, follow the path **S C**.

Now for a key question

Is S C a valid path in the S P C system?

Keep in mind as you try to answer this question that *it is as important to reflect upon your results as it is to answer the questions*. In order to reflect on what you have done, you need to record your efforts concisely and clearly. Before you begin, observe that *none of the rules are reversible*.

Example: P P P can be replaced with a C, but C cannot be replaced with P P P.

The S P C system is a well–kept secret, since not even all math teachers know about it. If you ask for help, be prepared to explain the rules. As you create valid production paths using the rules, see if you can answer Ernie's question about whether S C is a valid path in the S P C system.

Investigation

In an effort to determine if S C is a valid path in the S P C system, the employees derived several valid paths randomly to see what turned up.

8. Starting with S P, derive at least twenty different valid paths using the rules of the S P C system. You may apply the rules to any of the new valid paths you create, instead of always starting with S P. Record each derivation. Use clear, concise notation so that someone else can check your steps. Use separate paper and play with the system.

1. This system is based on the M U puzzle in Hofstadter, Douglas; *Gödel, Escher, Bach: An Eternal Golden Braid*; Vintage Books (New York, NY); 1980

9. Did you produce the same valid path by different derivations? Record one such valid path and the different derivations.

10. If you succeeded in producing S C, show your derivation. If not, write a conjecture about S C.

11. Describe any patterns you discovered.

Look for patterns.
Make and record your guesses.
Test your guesses.

Problem Solving Toolkit

Discussion

One of the important differences between humans and machines is that machines (such as calculators and computers) can be made to be totally unobservant. People cannot. One of the characteristics of intelligence is that it can jump out of the task that it is performing and survey what it has done. An intelligent person looks for and often finds patterns that make the job easier. Your calculator does not think for you. Effectively using the calculator and the computer in this course requires you to use your intelligence and analyze what you see.

Look back and reflect on what you have done.

Problem Solving Toolkit

Now that you tried solving the puzzle by generating valid paths, let's approach the problem mathematically. One way to look for patterns is to count the number of P's in each valid path. The number of P's in each valid path is called its ***P-count***. For example, S P P C P P C has an P–count of four.

Investigation

12. Look at each of the rules. Describe how each of the rules affects the P–count.

Make a table to help organize your work.

Problem Solving Toolkit

13. Complete Table 1. In the first column, record ten valid paths you produced in Investigation 8 that have **different** P–counts. Record the P–count for each valid path in the second column.

Valid Paths	P–Counts

Table 1: P–Counts of S P C valid paths

14. Compare Table 1 with others in your group. Can you think of numbers which are or could be P counts in your list? What numbers don't seem to appear? Write down your observations about the P–counts of the valid paths you have produced.

15. Use your observation about P–counts to determine whether or not S C is a valid path. List your reasons.

Discussion

In this last series of investigations, we analyzed P–counts. In fact, we have changed the task from one of producing the path S C to one of P–counts. S C is a valid path only if a P–count of 0 is possible. The question is changed from

Is S C a valid path in the S P C system?

to

Does a zero P–count exist?

Change your approach to the problem by asking a different, equivalent question.

Problem Solving Toolkit

Once a zero P–count becomes the goal of producing valid paths, we discover that

- P–counts increase by Rule 1.
- P–counts decrease by Rule 3.
- The P–count begins at 1 (1 is not a ***multiple*** of 3).
- Two of the rules do not affect the P–count at all.

Because a multiple of three does not exist initially, the two remaining rules which affect the P–count can never create a multiple of 3 and since 0 is a multiple of 3, the P–count can never be equal to 0. We conclude that

S C is not a valid path of the S P C system.

To a mathematician, "algebra" implies a mathematical structure in which the manipulation of symbols is only one feature. "Doing mathematics" includes the willingness to explore, to play with unfamiliar problems, even those which at first glance do not appear to have any connections with mathematics. We have discovered some of the mathematics in the S P C puzzle. This investigation is another example of the problem solving we began in Section 1.1. A key aspect of problem solving is the willingness to explore a problem situation. In doing so, we make mistakes. The process of exploring, making mistakes, and learning from these mistakes is a critical part of studying mathematics. Another crucial feature is reflecting on solutions. This includes being able to determine whether an answer is correct or not. Reflection also implies asking yourself if there might have been an easier way to solve the problem and whether there are connections between the given problems and others you have studied.

Explorations

1. Create your own personal dictionary of vocabulary words. List and define words in this section that appear in ***italics bold*** type using your own words.

2. Indicate which definitions you knew and which definitions you found in an outside source. List your source (i.e., dictionary, girlfriend, buddy, old math book, glossary, etc.)

3. Prove each of the following are valid paths in the S P C system by recording your derivations, using the axiom S P as a starting point for each derivation.

   a. S C P

   b. S P P P C P P C

   c. S P P P P P C C

4. Determine if the following are valid paths in the S P C system. If you think a combination of letters is a valid path, prove it is a valid path by using the rules and starting with S P record your derivation. If a combination of letters is not a valid path, explain why.

   a. S P C S

   b. S P P P P P C P P P P C P C

   c. S P P P C

   d. C S P P

   e. S P P C P P

   f. S C C C C C

5. Is there a test to detect **non–valid paths**? What might it be?

6. Which rules lengthen an existing valid path? Shorten an existing valid path?

7. How is working backwards different from the illegal reversal of the rules? Explain.

8. Explain why zero is a multiple of three.

9. Beginning with the production path S P P P P, if you apply Rule 1 is the new path S P P P P P P or S P P P P P P P P? Describe the difference between "doubling" and "counting by two".

10. List at least two ways that you worked within the system as you investigated the S P C system.

11. There is a list of standards for the use of notation (page 15). Give an example of notation which incorporates the standards. Give an example of notation which violates the standards.

12. List at least three ways that you worked outside the system as you investigated the S P C system.

13. What mistakes did you make as you explored the S P C system? How did you discover you had made a mistake (book, teacher, friend, etc.)? What did you learn from this?

14. Does this seem like mathematics to you? Why or why not?

15. Write down your feelings about the following inquiries.

    As you produced valid paths did you

    a. organize what you had produced?

    b. try to figure out what you had missed?

    c. start getting anxious?

    d. wonder if you could succeed in producing S C?

    e. think it was fun trying to determine if S C was a valid path?

Concept Map

Construct a concept map centered on the **S P C system**.

Reflection

Summarize the process you used to investigate the S P C system using the concept map as a guide. How did you get started? Describe all subsequent tasks in your investigation, using complete sentences.

Section 1.3 Using Variables to Generalize

Purpose

- Introduce domain of a relationship.
- Use variables to generalize patterns and rules.
- Explore structure through the use of variables.

Investigation

We begin this section by identifying some specific features of paths and rules in the S P C system.

1. Given the path S P P P P, state which rules can be applied to obtain a new path. Defend your decision.

2. Given the path S P P C C P P C P C, state which rules can be applied to obtain a new path. Defend your decision.

3. What must be true about a path before you can apply

   a. Rule 1?

   b. Rule 2?

   c. Rule 3?

   d. Rule 4?

4. Draw a box around the portion of the path that is not important when deciding if you can apply the given rule. Do not include the start–up S in the box.

   Example: To apply Rule 2 to S P P P P, the only letter that is important is the last P since the path must end in P to apply Rule 2. So draw a box around the first three P's.

   a. Apply Rule 1 to S P P P P.

   b. Apply Rule 1 to S P P C C P P C P C.

   c. Apply Rule 2 to S C P P C P P.

Discussion

Investigations 1–4 are examples of working outside the S P C system. Instead of just deriving paths, we are observing important connections between paths and rules.

When we determine what must be true about a path before a rule is applied, we are finding the ***domain*** of the rule.

The domain of Rule 1 is the set of all paths since Rule 1 can be applied to every legal path.

The domain of Rule 2 is the set of all paths that end in P.

The domain of Rule 3 is the set of all paths that contain three consecutive P's.

The domain of Rule 4 is the set of all paths that contain two consecutive C's.

We have used notation in the S P C system to record what was done. If we agree on notation that allows us to rewrite the rules of the system, it is easier to know what to do and how to apply the rules to an existing path or to the axiom. Changing a rule from words to notation makes the rule easier to understand, provided the notation makes sense. You change a rule from words to notation one of two ways:

a. rewrite the words in the rules in terms of symbols or

b. look at examples of the rules being used and generalize what you observe using ***variables***.

In Investigation 4 you observed that the combination of letters in the boxes was not important in deciding if a given rule could be applied.

Apply Rule 1 to S P P P P

Apply Rule 1 to S P P C C P P C P C

Apply Rule 2 to S C P P C P P

S did not change when Rule 1 was applied, but the sequence of letters following the letter S was repeated. Looking at one example is not enough information on which to form a generalization, but, based on your investigations in the preceding section and this one, if you draw a box around the four P's following the letter S, a general form for Rule 1 is suggested.

S P P P P $\xrightarrow{\text{Rule 1}}$ S P P P P P P P P

In general, we could express Rule 1 as

S □ $\xrightarrow{\text{Rule 1}}$ S □ □

The box on the left can be filled with the combination of P's and C's of any existing path, but then every box on the right must be filled with that same sequence of letters. Since what you place in the left box may vary, it is called a ***variable***. Instead of a box, mathematicians usually use letters as variables to generalize a rule because letters are easy to write and to typeset. Using the letter x, Rule 1 is written as

$$\text{S}x \xrightarrow{\text{Rule 1}} \text{S}xx$$

Now that we have a generalized form, we can apply Rule 1 to any path more easily. In S P P C C C, the sequence P P C C C would be placed in each box or would replace every x.

Let's investigate the generalization of Rules 2–4.

Investigation

5. Apply Rule 2 to the following paths. In both cases, draw a box around the portion of the path that is not important when deciding if you can apply Rule 2. The box should appear around the same combination of letters both before and after applying the rule.

   a. S P P P P

   b. S C P P C P P

6. a. Generalize Rule 2 using boxes.

   b. Generalize Rule 2 using a variable.

7. What would the variable represent if Rule 2 is applied to the path S C C P P?

8. A path must have three consecutive P's to apply Rule 3. Draw a box around the portion of the path that is between the S and the three P's. Draw a circle around the portion of the path that follows the three P's.

   Example: To apply Rule 3 to S P P C P P P C, the three P's are preceded by P P C. So draw a box around P P C. The three P's are followed by C. Draw a circle around the C.

   S [P P C] P P P (C)

   a. S C P P P C P P

   b. S P C P P P

   c. S P P P C P P C

9. Apply Rule 3 to the following production paths. In each case, draw a box around the portion of the path that precedes the three P's and a circle around the portion of the path following the three P's. The box and circle should appear around the same combination of letters both before and after applying the rule.

   a. S C P P P C P P

   b. S P C P P P

   c. S P P P C P P C

10. a. Generalize Rule 3 using boxes and circles.

b. Generalize Rule 3 using variables x and y.

11. A path must have two consecutive C's to apply Rule 4. Draw a box around the portion of the path that is between the S and the two C's. Draw a circle around the portion of the path that follows the two C's.

Example: To apply Rule 4 to S P P C C P P C, the two C's are preceded by P P. So draw a box around P P. The two C's are followed by P P C. Draw a circle around P P C

a. S C P P C C P P

b. S C C P P P

c. S P P P C P C C

12. Apply Rule 4 to the following paths. In each case, draw a box around the portion of the path that precedes the two C's and a circle around the portion of the path following the two C's. The box and circle should appear around the same combination of letters both before and after applying the rule.

a. S C P P C C P P

b.S C C P P P

c. S P P P C P C C

13. a. Generalize Rule 4 using boxes and circles.

b. Generalize Rule 4 using a variables x and y.

Discussion

In order to generalize the other three rules, we must precisely determine what the variable, (the symbol $\square$ or x), represents. This is true every time you use a variable, not only in the S P C system, but also in mathematics, or anywhere else. State what the variable represents in writing before solving a problem.

To determine what the variable represents, we look at the restrictions required for each rule. Since we discovered that every rule begins with S, each generalization also begins with S. The restriction on Rule 2 is that the path must end in P. The sequence of letters between S and the last P will vary, thus we generalize Rule 2 by replacing the sequence of letters between S and the last letter, P, with a variable.

This is demonstrated in the following answers to Investigation 5.

a. $$\text{S}\boxed{\text{P P P}}\text{P} \xrightarrow[\text{Rule 2}]{} \text{S}\boxed{\text{P P P}}\text{P C}$$

b. $$\text{S}\boxed{\text{C P P C P}}\text{P} \xrightarrow[\text{Rule 2}]{} \text{S}\boxed{\text{C P P C P}}\text{P C}$$

Rule 2 can be expressed as

$$\text{S}\boxed{}\text{P} \xrightarrow[\text{Rule 2}]{} \text{S}\boxed{}\text{P C}$$

or, using a letter as the variable,

$$\text{S}\,x\,\text{P} \xrightarrow[\text{Rule 2}]{} \text{S}\,x\,\text{P C}$$

The restriction on Rule 3 is that there must be three P's together. The letters between the S and the three P's and the letters after the three P's are not important to the application of Rule 3, but since the two collections may be different sequences of letters, we need to use two different symbols, or variables, to generalize Rule 3. One variable cannot represent two different objects (sequences of letters) in the same problem.

For example, the answers to Investigation 9a. and b. are

a. $$\text{S}\boxed{\text{C}}\text{P P P}\enclose{circle}{\text{C P P}} \xrightarrow[\text{Rule 3}]{} \text{S}\boxed{\text{C}}\text{C}\enclose{circle}{\text{C P P}}$$

b. $$\text{S}\boxed{\text{P C}}\text{P P P} \xrightarrow[\text{Rule 3}]{} \text{S}\boxed{\text{P C}}\text{C}$$

Note that the circle is empty in Investigation 9b. The box would be empty in Investigation 9c.

This leads to the generalization of Rule 3

$$S\square PPP\bigcirc \xrightarrow[\text{Rule 3}]{} S\square C\bigcirc$$

or

$$S\,x\,PPP\,y \xrightarrow[\text{Rule 3}]{} S\,x\,C\,y$$

The restriction on Rule 4 is that there must be two C's together. The letters between the S and the two C's and the letters after the two C's are not important to the application of Rule 4. As in Rule 3, the sequence before the two C's and the sequence after the two C's may be different.

The answers to Investigation 12a. and b. are

a. $S\boxed{CPP}CC\,(PP) \xrightarrow[\text{Rule 4}]{} S\boxed{CPP}\,(PP)$

b. $SCC\,(PPP) \xrightarrow[\text{Rule 4}]{} S\,(PPP)$

Note that the box is empty in Investigation 12b. The circle would be empty in Investigation 12c.

This leads to the generalization of Rule 4.

$$S\square CC\bigcirc \xrightarrow[\text{Rule 4}]{} S\square\bigcirc$$

or

$$S\,x\,CC\,y \xrightarrow[\text{Rule43}]{} S\,x\,y$$

We have used variables to generalize the rules of the S P C system. In this case, the variables represented the combination of letters

- that were unimportant in deciding if a rule could be applied and,
- that remained constant when the rule was applied.

We summarize the generalization of all the rules in Figure 1.

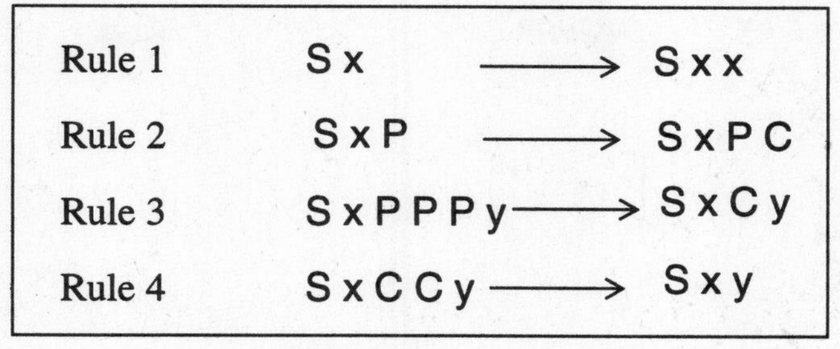

Rule 1	S x	⟶	S x x
Rule 2	S x P	⟶	S x P C
Rule 3	S x P P P y	⟶	S x C y
Rule 4	S x C C y	⟶	S x y

Figure 1

Now we will look at variables as generalizers of arithmetic statements.

Investigation

14. Calculate the following. Verify your answers on a calculator.

a. 5 + 0 b. 5 x 0

c. –91 + 0 d. 2.47 + 0

e. (4 + 8 – 2) + 0 f. –91 x 0

g. 2.47 x 1 h. (4 + 8 – 2) x 1

i. 2.47 x 0 j. (4 + 8 – 2) x 0

15. Identify the problems in Investigation 14 that have a common property.

16. For each different property identified:

a. List the problems with that common property.

b. Describe each common property in writing.

17. Generalize each of the calculations in Investigation 14 using a variable to rewrite the problem.

18. Let the variables n and d represent the whole number numerator and denominator of any fraction. Use n and d to write the general form of a fraction.

Discussion

The power of mathematics comes from our ability to generalize several examples like $5 \cdot 1 = 5$ to the statement:

For any number n, $n \cdot 1 = n$

We often name these generalized statements. The example above, $n \cdot 1 = n$, is called the ***identity property for multiplication***. In investigation 12 you wrote generalized statements of the identity property of addition and the zero property of multiplication.

It is important to be aware of the assumptions we make when we are using variables.

Assumptions about variables:

- In a problem or statement, every time a variable is used, it represents the same thing.
- The variable probably represents something different when we work a new problem.
- Variables make it possible to describe a few specific examples as a generalized statement.
- A generalized statement must always be true and, in order to guarantee this, the use of restrictions, or conditions, may be required. For example, $\frac{n}{d}$ represents any fraction only if $d \neq 0$.
- Variables allow us to save a lot of manual effort since a single variable can represent several words or a tedious repetition. The cost of using a variable to simplify our work is increased mental effort needed to deal with the abstraction.

Explorations

1. List and define words in this section that appear in ***italics bold*** type using your own words.

2. Indicate which definitions you knew and which definitions you found in an outside source. Indicate your source.

3. Using Rule 1, draw a box around the sequence of letters that can be represented by the variable in each of these paths.

   a. S P
   b. S C P
   c. S P P P P C P P P P C
   d. S C P P C

4. Using Rule 2, draw a box around the sequence of letters that can be represented by the variable in each of these paths.

   a. S P
   b. S C P
   c. S P P P P C P P P P C
   d. S C P P C

5. Using Rule 3, draw a box and a circle around the sequences of letters that can be represented by the variables in each of these paths.

   a. S P C P P P P
   b. S P P P P C P
   c. S P P P P
   d. S P P P P C C P C P P

6. Using Rule 4, draw a box and a circle around the sequences of letters that can be represented by the variables in each of these paths.

   a. S P C C P
   b. S C C P P C
   c. S P P P P P P P P
   d. S P P P C C P C P P

7. List five specific numerical examples for each of the following generalizations.

   a. $1 \cdot n = n$

   b. $a \cdot a \cdot a = a^3$

   c. $\sqrt{a^2 + b^2} \leq a + b$, a and b are whole numbers

8. Give an example of

a. a noun

b. a noun which ends in **s**

c. a noun that ends in **x**

d. a verb

9. Write down what each variable represents and then use the variable to write the grammar rule for

a. making most nouns plural.

b. making a noun that ends in **s** plural.

c. making a noun that ends in **x** plural.

d. making some verbs past tense.

10. Do English grammar rules work as nicely as those in the S P C system? Explain.

11. a. Use your calculator to determine the answers.

1^2 =

11^2 =

111^2 =

b. Predict and check

1111^2 =

11111^2 =

c. Since the answer to 111111^2 is displayed in scientific notation, all the digits you would get if you multiplied this by hand will not be visible. Predict exact answers to the following.

111111^2 =

1111111^2 =

111111111111^2 =

d. Check the last answer on a scientific or graphing calculator. Does your generalization approximate the calculator's answer?

e. Multiply 111111111111^2 out by hand and check your answer. If you are correct, explain in words why your generalization worked. If not tell what restriction in the problem you overlooked.

12. John was running late to class and parked illegally in the faculty lot on Monday morning at 10:05. He returned at 10:55 and was elated to find no ticket. A week later he was really late and at 10:15 parked there again. He was pleasantly surprised to find no ticket. The third Monday, he arrived at 9:50 and was unconcerned as he parked in the faculty lot. What is John's conjecture? Explain the probable implications of John's generalization.

13. For each of the five assumptions about variables (page 32) state the assumption and give an example that demonstrates your understanding of it.

Concept Map

Construct a concept map centered on **variable**. Include your prior experiences with variable as well as the experiences developed in this text.

Reflection

If you could do something to make the concept of variable more understandable for a friend, what would you do?

Section 1.4 Expanding the Notion of Variable

Purpose

- Explore more uses for variables.
- Compare arithmetic representations with algebraic representations.

Discussion

We have used variables to generalize rules and properties of numerical computation. The ability to generalize using variables is a major objective for students studying algebra. You have already done this many times before—perhaps without thinking about why. The following investigation focuses on how mathematics uses symbols to indicate both what you do (process) and the answer you get (output).

Investigation

1. For each of the following numerical expressions, describe in words the process (what you are supposed to do) and the output (the answer you get when you do it).

Expression	Process	Output
$50 - 9$		
$14 + 5$		
$2\,(5)$		
5^2		

Table 1: Numerical process and output

For each of the following variable expressions describe the process and the output.

Expression	Process	Output
$50 - p$		
$n + m$		
$2n$		
n^2		

Table 2: Variable process and output

2. Compare the output of $50 - 9$ with the output of $50 - p$. Describe your observations.

3. If I have $50 and want to buy a book that costs $9, what does $50 - 9$ represent?

4. If I am considering one of several different books all with different prices what does $50 - p$ represent?

5. What numerical limitations must be placed on the value of p if 50 represents the amount of money I have and p represents the price of a book?

6. Describe, in writing, a situation that could be represented by

   a. 2×5

   b. $2n$

7. a. Complete Table 3 by choosing six different values for n larger than one.

n, n larger than one	**$2n$**

Table 3: Doubling investigation 1

b. Based on the results, which is larger: n or $2n$ if n is larger than one? Defend your answer.

8. a. Complete Table 4 by choosing six different values for n between zero and one.

n, n between zero and one	$2n$

Table 4: Doubling investigation 2

b. Based on the results, which is larger: n or $2n$ if n is between zero and one? Defend your answer.

Discussion

Did you notice that, when using variables, the expression itself is often the output? Thinking of the output $50 - p$ as "my change" may help you distinguish it from the process of subtracting the price of the book from $50. However, when the symbols are not given in a context, it is sometimes difficult to make a distinction between *what to do* (the process) and *the answer* (output), both of which are written and look the same. As you continue to study mathematics, it is important that you learn to choose the appropriate use of an expression as either process or output.

When using variables in an expression, it is important to consider the values that may be used when substituting for the variable. The set of all values that may be used for the variable is often called the ***domain*** of the variable. The p in the expression $50 - p$ can take on all numbers with two decimal places between 0 and 50 inclusive—this is the domain of variable p for the given problem situation. If p were smaller than 0, it would represent a negative price for a book. If p were larger than 50, then the expression $50 - p$ would be negative. That means I could not afford that book!

When substituting values for the variable n in the expression $2n$, the domain for n is the set of all numbers. From the investigation, note that regardless of the positive number substituted for n, $2n$ is larger than n.

In our consideration of $50 - p$ and $2n$, we were using the variable as an unknown number. Once a number in the domain of the variable is selected, the number replaces the variable and the expression has one unique output.

Let's continue using variables to further investigate the P–counts in S P C legal paths. Once we know that the P–count of a legal path in the S P C system cannot be a multiple of three, another question occurs. Can the P–count be *any* whole number except a multiple of three?

The only rules that change the P–count of a legal path are Rule 1 and Rule 3. Let's investigate the P–counts when Rule 1 and Rule 3 are applied repeatedly beginning with the initial path S P.

Investigation

9. Begin with the initial path S P and repeatedly apply only Rule 1. Record the results in Table 5.

Number of times Rule 1 is applied	P–count
1	
2	
3	
4	
5	
6	
7	
8	

Table 5: P–counts using only Rule 1

10. What property do the P–counts have if only Rule 1 is applied?

11. a. If Rule 1 is applied 20 times, what is the P–count? How did you determine the P-count?

b. If Rule 1 is applied n times, write an expression containing the variable n that represents the P–count. Why is your expression a representation of the P-count?

12. S P P P P P P P P P P P P P P P P P P P P P P P P P P P P P P P P was derived from the initial path S P by applying Rule 1 five times.

a. Complete Table 6 in which Rule 3 is successively applied to the given path.

Number of times Rule 3 is applied	Total number of P's removed from the theorem	Number of P's remaining in the theorem.
0	0	32
1	3	
2		
3		
4		
5		
6		

Table 6: P–counts using Rule 3

b. If Rule 3 is applied nine times, what is the total number of P's removed? How did you determine the total number of P's?

c. If Rule 3 is applied x times, write an expression containing the variable x for the total number of P's removed. Why is your expression a representation of the total number of P's removed?

d. If Rule 3 is applied nine times, what is the total number of P's remaining in the path? How did you determine the number of P's remaining?

e. If Rule 3 is applied x times, write an expression containing the variable x for the number of P's remaining in the path. Why is your expression for the number of remaining P's correct?

f. What is the domain of the variable x for the given path? Justify your answer.

Discussion

Beginning with the initial path S P, applying Rule 1 five times results in a P–count of 2^5 or 32 P's. Applying Rule 1 ten times results in 2^{10} or 1024 P's. In general, applying Rule 1 n times results in 2^n P's. The expression 2^n is an example of an algebraic expression that generalizes the P–count for the given situation. In this case the variable n is used to generalize a pattern.

After applying Rule 1 five times beginning with the initial path S P, we can then repeatedly apply Rule 3 to reduce the P–count. If Rule 3 is applied two times, then $3\,(2)$ or 6 P's are removed. If Rule 3 is applied five times, then $3\,(5)$ or 15 P's are removed. Generalizing, if Rule 3 is applied x times, then $3\,(x)$ P's are removed. The expression $3\,(x)$ is another example of an algebraic expression—this expression generalizes the number of P's removed by successively applying Rule 3. Again the variable x is used to generalize a pattern. The value of the variable x is restricted by the number of times three successive P's can be removed from a legal path. In the case of the legal path containing an S followed by thirty–two P's, Rule 3 can be applied at most ten times. The domain of the variable x in this case is the set of whole numbers from 0 to 10 inclusive.

Now let's put the two ideas together to write an expression that represents the P–count as determined by how many times Rule 1 is applied to the initial path S P and by the number of times Rule 3 is applied.

Investigation

13. Start with the initial path S P. If Rule 1 is applied twenty times and Rule 3 is applied thirty–two times, write a *numerical expression* representing the process required to find the number of P's remaining in the path. Explain why your process is appropriate.

14. Start with the initial path S P. If Rule 1 is applied n times and Rule 3 is applied x times, write an *algebraic expression* containing the variables n and x for the number of P's remaining in the path. Explain how you determined that this expression is correct.

Discussion

If Rule 1 is applied twenty times and Rule 3 is applied thirty–two times, the P–count of the resulting path is $2^{20} - 3(32)$ or 1,048,480. In general, if Rule 1 is applied n times and Rule 3 is applied x times, the P–count of the resulting path is represented by the expression

$$2^n - 3x.$$

We have discovered a relationship between three quantities: the number of times Rule 1 is applied; the number of times Rule 3 is applied; and, the P–count of the resulting path. As we've derived it, the P–count is dependent on the other two quantities. Many such relationships are represented algebraically using ***equations***. In this situation, the equation would be

$$P\text{–}count = 2^n - 3x.$$

Equations, such as the one above, that describe relationships are a focus of this text.

If we look at a dictionary, we might find ***arithmetic*** defined as the branch of mathematics concerned with numerical calculations. When we completed the tables associated with repeatedly applying Rule 1 and repeatedly applying Rule 3, we performed numerical calculations. This is an example of applying arithmetic to the problem of determining P–counts. On the other hand, when we used variables to generalize a pattern, we used algebra to represent the computation of P–counts.

In order to establish the definitions of arithmetic and of algebra, it is helpful to look at how problems are solved using arithmetic or using algebra. Comparing solutions using arithmetic with those using algebra reveals how problems are alike and how they are different. By solution, we mean the process that is used as well as the answer. One way to classify a solution as algebraic or arithmetic is to determine whether the problem can be solved using computation only, or if variables were used to represent the problem situation. Investigating and comparing what happens when you work several different problems develops understanding about what is arithmetic and what is algebra.

Explorations

1. List and define words in this section that appear in ***italics bold*** type using your own words.

2. Review the parking lot problem discussed in Chapter 1, Investigation 1 (page 1). Several solution techniques are suggested. Identify the techniques that are arithmetic. Identify the techniques that are algebraic. Justify your answer.

3. a. Complete Table 7 by choosing six different values for n larger than one.

n, n larger than one	n^2

Table 7: Squaring investigation 1

b. If n is larger than one, based on the results, which is larger: n or n^2 ? Defend your answer.

4. a. Complete Table 8 by choosing six different values for n between zero and one.

***n*, *n* between zero and one**	n^2

Table 8: Squaring investigation 2

b. If n is between zero and one, based on the results, which is larger: n or n^2? Defend your answer.

5. Describe the difference between the expressions

a. $2n$ and n^2. Discuss the process for each.

b. n^2 and 2^n. Discuss the process for each.

6. Substitute five different positive values for p in the expression $50 - p$. Based on the results, how does the size of p affect the size of $50 - p$?

7. Given the P–count equation *P–count* $= 2^n - 3x$. Assume we start with the initial path S P.

a. Find the P–count if Rule 1 is applied seven times and Rule 3 is applied eleven times.

b. Find the number of times rule 3 was applied if Rule 1 was applied four times and the P–count is seven.

c. Find the number of times Rule 1 was applied if Rule 3 was applied six times and the P–count is 46.

d. Find all possible P–counts if Rule 1 was applied four times.

8. How many times should Rule 1 be applied and Rule 3 be applied to obtain a P–count of:

   a. 13?

   b. 52?

   c. Did you find the answer to part b numerically or algebraically? Explain.

9. What is the domain of the variable x in the P–count equation if Rule 1 is applied six times? Justify your answer.

10. Describe the ways that a variable was used in this section.

11. Let the letter n represent the number of letters following the start–up S in a given path in the S P C system.

    a. Use n to write an expression that would represent the number of letters following the S in the path after Rule 1 is applied.

    b. Use n to write an expression that would represent the number of letters following the S in the path after Rule 2 is applied.

    c. Use n to write an expression that would represent the number of letters following the S in the path after Rule 3 is applied.

    d. Use n to write an expression that would represent the number of letters following the S in the path after Rule 4 is applied.

Concept Map

Construct a concept map centered around the word **algebra**.

Reflection

Discuss your idea of arithmetic and your idea of algebra. Write a paragraph, no more than one–half page in length, using complete sentences.

Section 1.5 Making Connections: What does it mean to do Mathematics?

Purpose

- Reflect upon ideas explored in Chapter 1.
- Explore the connection between doing math and understanding.

Investigation

In this section you work outside the system to reflect upon the mathematics in Chapter 1: what you've done and how you've done it.

1. State the five most important ideas in this chapter. Why did you select each?

2. Identify all the mathematical concepts, processes, and skills you used to investigate the problems in Chapter 1.

3. What has been common to all of the investigations which you have completed?

4. You have investigated many problems in this chapter.

   a. List your three favorite problems and tell why you selected them.

   b. Which problem did you think was the most difficult?

5. When doing mathematics, is the number of problems completed an important factor? Why or why not?

6. Select a key idea from this chapter. Write a paragraph explaining it to a confused best friend.

Discussion

There are a number of really important ideas which you might have listed. Some of these include variable, domain, problem solving heuristics, multiple solutions, patterns, the development of notation, process vs answer, exploring problems within the system and reflecting upon problems outside the system. You have also been introduced to the idea of proof.

Concept Map

Construct a concept map centered around the phrase **doing mathematics**.

Reflection

Before beginning this class, what was your idea of "doing mathematics"? How has your work in Chapter one altered this idea?

Illustration

Draw a picture of **a mathematician**.

THE
MATHEMATICIAN

Chapter 2

Whole Numbers: Introducing a Mathematical System

Section 2.1 Whole Number Domains

Purpose

- Explore the domain of problem situations.
- Introduce sets in the context of the problem domain.
- Identify common characteristics of sets.
- Develop the connection between the S P C system and the system of whole numbers, another mathematical system.

Investigation

The parking lot problem from Section 1.1 (Investigation 1) used variables for the number of cars and for the number of motorcycles. The parking lot had room for twenty vehicles and was assumed to be full. For the next investigation, use the same parking lot, but it is not necessarily full. Instead of considering the number of cars or the number of motorcycles, let's use one variable, w, to represent the number of wheels parked in the lot at any given time.

1. a. If only cars could park in the lot, list the domain for variable w, the number of wheels, that might be in the parking lot at any given time.

   b. Describe the set of numbers listed as an answer to part a in words.

2. a. If only motorcycles could park in the lot, list the domain for w, the number of wheels, that might be in the parking lot at any given time.

   b. Describe the set of numbers listed as an answer to part a in words.

3. a. If both cars and motorcycles are allowed to park in the lot, list the domain for w, the number of wheels, that might be in the parking lot at any given time.

   b. Describe the set of numbers listed as an answer to part a in words.

Consider the P–counts in the S P C system. The formula developed in Section 1.4 is

$$P\text{–}count = 2^n - 3x$$

where n is the number of times Rule 1 is applied and x is the number of times Rule 3 is applied. Assume that we begin with the initial path S P.

4. List the set of numbers that acts as the domain for the variable n. Justify your answer.

5. Assume n has a value of six.

   a. List the set of numbers that acts as the domain for the variable x. Justify your answer.

   b. List the set of numbers that acts as the set of possible values for *P–count*. Justify your answer.

6. Describe in words the possible values of the *P–count* regardless of the choice of n and x.

Discussion

It is important to think about the type of number that is appropriate as an answer to a given problem. For example, a fraction is appropriate to describe the portion of a pizza I devoured, but is not appropriate to describe how many calculators I own.

The collection of numbers that can be used as possible replacements for a variable in a given problem is called the ***domain*** of the problem situation. Domains are represented mathematically using ***sets***. The answers to the previous investigations are sets consisting of whole numbers. The phrase, "A ***set*** of objects", is part of the language of mathematics. Objects are included in a set because they have common characteristics. The whole numbers are an example of a set. The set of whole numbers, like any set, must be ***well–defined***, i.e., it must be possible to determine if a number is, or is not, a whole number.

We can represent the set of whole numbers by a number line graph (See Figure 1),

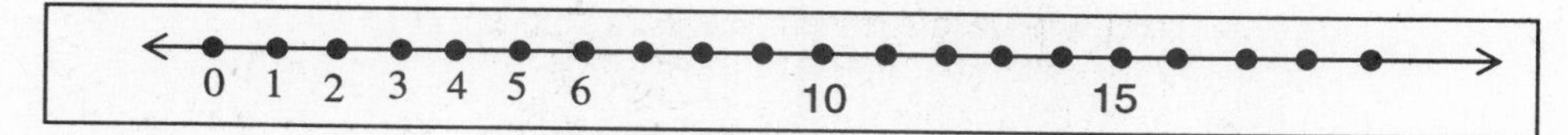

Figure 1

or by a listing

$$W = \{0, 1, 2, 3, 4, ...\}$$

W is a variable name representing the set of whole numbers. The ***elements*** of the set are enclosed in ***braces***, **{ }**, and are separated by commas. Three dots, called an ***ellipsis***, must be used as the last element when the whole numbers are listed, since there is no largest whole number. The set of whole numbers is an example of an ***infinite set***, i.e., it has no last element. Any collection of numbers that contains only some whole numbers is called a ***subset*** of the whole numbers.

If only automobiles are allowed in the parking lot, then the domain set for the number of possible wheels in the parking lot at any given time is

$$\{0, 4, 8, 12, 16, 20, 24, 28, 32, 36, 40, 44, 48, 52, 56, 60, 64, 68, 72, 76, 80\}$$

This is the set of whole number multiples of four less than 81. It is a subset of the whole numbers. It is a ***finite*** subset since it contains only twenty–one elements. This set is written in abbreviated notation by listing the first few elements of the set establishing the sequence and using ellipsis to indicate a continuation of the sequence. For example,

$$\{0, 4, 8, 12, ..., 80\}$$

Another finite subset of the whole numbers is the domain set for the number of wheels if only motorcycles are allowed to park in the lot. In this case, the set is

$$\{0, 2, 4, 6, 8, \ldots, 40\}$$

This is the set of even whole numbers less than 41.

If both types of vehicles are allowed to park in the lot, numbers that were not in either of the previous sets form the domain, which is the set

$$\{0, 2, 4, 6, 8, \ldots, 80\}.$$

With respect to the P–counts of paths in the S P C system, we have more subsets of the whole numbers. Since Rule 1 can be applied as many times as we wish, the domain set of n, the number of times Rule 1 is applied, is the set of whole numbers. If n has a value of six, then the path has 64 P's. The domain set for x, the number of times Rule 3 is applied, is restricted to the whole numbers less than or equal to $\frac{64}{3}$.

This domain set for variable x is $\{0, 1, 2, 3, 4, \ldots, 21\}$.

This implies that the number of P's removed is the set $\{0, 3, 6, 9, \ldots, 63\}$.

Thus, given sixty–four P's, the possible values for *P–count* in this case is $\{64, 61, 58, 55, 52, 49, \ldots, 4, 1\}$.

Finally, recall that the P–count must not be a multiple of three. So the set of possible values for P–counts, in general, is the set of whole numbers that are not multiples of three.

Investigation

A very important infinite subset of the whole numbers is the set of ***prime numbers***. This investigation provides an opportunity to explore this set.

7. A ***prime number sieve*** is a means of identifying prime numbers. To find the prime numbers, do the following:

a. Complete Table 1 for the numbers up to and including 103.

2	3	4	5	6	7
8	9	10	11	12	13
14	15	16	17	18	19
20	21	22	23	24	25
26	27	28	29	30	31
32	33	34	35	36	37
38	39	40			

Table 1: Prime number sieve

b. Circle the first number in Table 1 and then cross out all multiples of that number to the end of the table.

c. Starting at the beginning of Table 1, circle the first number that is not already circled or crossed out. Cross out all multiples of that number.

d. Repeat steps c and d until all numbers are circled or crossed out.

e. Record all the circled numbers. These are the prime numbers less than or equal to 103.

8. Are there any patterns in Table 1 (page 53)? If so, describe those you observe.

9. What was the largest number you circled that still had multiples in the table that were not crossed–out?

10. Write your definition of a prime number. Compare your definition with the definitions of your group members.

11. Test your definition on the numbers 2, 4, 51, 109, and 121. Which are prime?

12. Is there a whole number between **6** and **7**? Justify your answer using a number line.

13. Is there a smallest whole number? If so, what is it? Justify your answer using a number line.

14. For each of the following numbers, determine whether the number is or is not a whole number. Give a reason for each answer.

a. $\frac{1}{2}$ b. -5

c. $\sqrt{4}$ d. $\sqrt{5}$

e. $\frac{12}{4}$ f. 0

g. 1 h. 0.5

i. .333 . . . j. $-\frac{2}{7}$

Discussion

A ***prime number*** is a whole number with exactly two distinct divisors, one and the number itself. In the prime number sieve, a number that is crossed out could not be prime since it is a multiple of some previous number in the table—thus the number had a divisor other than one and the number itself.

Non–zero whole numbers that have more than two divisors are called ***composite numbers***. Thus, using these distinctions, the whole numbers can be ***partitioned*** into three distinct sets: the prime numbers; the composite numbers; and, the set {0, 1}. The union of these three sets is the set of whole numbers.

In order to cross out all the composite numbers in Table 1 (page 53), you needed only to cross out the multiples of the primes 2, 3, 5, and 7. The remaining numbers in the table that have not been crossed out are prime. In general, to test a number for primality, divide the number by all primes starting at two until the square of the prime exceeds the number. If no previous prime is a divisor of the number, the number is prime.

For example, the largest number in our prime number sieve is 103. The square of seven is forty–nine. The square of eleven, the next prime, is 121, which exceeds 103. By the time you eliminate the multiples of the primes up through seven, you have eliminated all composites. You need to think about why this works.

Let's discuss the remaining investigations. The set of whole numbers between 6 and 7 is also a subset of the whole numbers. It is a very special set. Since it contains no elements it is called an ***empty set*** and is represented by either the symbol ∅ or the set notation { }.

Working with sets provides another way to use a variable. We could select your favorite letter, such as *n*, to represent any whole number. Whenever we see *n* later, we may think: "That's a whole number!"

If we wish to express the phrase "7 more than a whole number", we can write $n + 7$, using the variable *n* to represent any whole number in a given set.

In Investigation 14 (page 55), note that the numbers 0 and 1 are whole numbers. If we ***simplify*** $\sqrt{4}$ we get 2, a whole number. If we simplify $\frac{12}{4}$, we get 3, so both $\sqrt{4}$ and $\frac{12}{4}$ are classified as whole numbers. The set of whole numbers is a subset of the set of real numbers. The other numbers in the list are not whole numbers, but are elements in the set of real numbers.

Investigation

15. What operations are used on the whole numbers?

16. Does the use of each of these operations always produce an answer that is another whole number? Clearly explain your answer.

Discussion

The set of whole numbers, along with the operations of addition, subtraction, multiplication, or division, some axioms, definitions, and theorems, form a mathematical system. An operation is ***closed*** within a system if the operation returns an answer that is an element within the system. For example, the number **4** and the number **3** are elements within the whole number system. Their sum, **7**, is also a whole number. If we add any two whole numbers, we get a whole number for the sum.

We say that addition is a closed operation on the whole number system.

However, though **3** is a whole number and **5** is also a whole number, when you subtract 5 from 3 the answer is **–2**, not a whole number, i.e., $3 - 5 = -2$. We say that subtraction is not a closed operation on the whole number system.

Similarly, multiplication is a closed operation on the whole numbers since the product of two whole numbers is always a whole number. Division is not a closed operation on the whole numbers since the quotient of two whole numbers is often not a whole number. The quotient of $1 \div 2$ is a good example.

Since the operations of addition and multiplication on whole numbers always return an answer that is also a whole number, we say that the set of whole numbers has ***closure*** for the operations of addition and multiplication. The operations of subtraction and division do not always return an answer that is a whole number, so the set of whole numbers is ***not closed*** under the operations of subtraction and division.

The fact that the whole number system is not closed under both subtraction and division is a reason we will explore other number systems later in the text.

Explorations

1. List and define words in this section that appear in ***italics bold*** type using your own words.

2. Indicate which definitions you knew and which definitions you found in an outside source. Indicate your source.

3. List the elements in each of the following subsets of the whole numbers.

   a. The set of odd numbers.

   b. The set of even numbers.

   c. The set of digits.

   d. The set of whole numbers that are their own squares.

   e. The set of prime numbers.

   f. The set of whole numbers between 11 and 12.

   g. The set of composite numbers.

   h. The set of numbers that are neither prime nor composite.

4. For each set in Exploration 3, describe in complete sentences the common characteristics of each set.

5. For each set in Exploration 3, record whether the set is infinite or finite. Justify your answers.

6. For each set in Exploration 3, determine if the set is closed under the operations of addition or multiplication.

7. If only automobiles are allowed in the parking lot, the possible number of wheels in the parking lot was expressed as the set of whole number multiples of four less than 81. Why was the number 81 used? What other numbers could have been used in its place?

8. Recall the expression $2^n - 3x$ for P–counts in the S P C system. The domain set for x is restricted to the whole numbers less than or equal to $\frac{64}{3}$ when n had a value of six. Why was $\frac{64}{3}$ used?

9. This question refers to the S P C system. List or describe the elements of each of the following sets and determine if the set is finite or infinite.

   a. The set of production steps.

   b. The set of initial paths.

   c. The set of rules.

   d. The set of paths.

   e. The set of valid paths

   f. The set of paths that do not begin with a start–up.

10. Given the set of legal paths and the set of paths in the S P C system, is either a subset of the other? Defend your answer in complete sentences.

11. Write a sentence describing each of the following sets.

    a. $\{0, 1\}$

    b. $\{1, 3, 5, 7, 9\}$

    c. $\{0, 1, 4, 9, 16, 25, \ldots\}$

    d. $\{0, 7, 14, 21, 28, \ldots\}$

12. Explain why 0 and 1 are not prime numbers.

13. a. Is 287 a prime number? Justify.

    b. Is 283 a prime number? Justify.

    c. What is the largest prime that must be used to test 287 or 283 for primality? Why?

14. Let the letter m represent any whole number. Write an expression for the following whole numbers.

    a. The whole number that is 9 larger than m.

    b. The whole number that is 4 smaller than m.

    c. The whole number that is twice m.

    d. The whole number that is 3 more than twice m.

    e. The next two whole numbers following m.

15. Represent each quantity with a variable. Indicate which variables have a domain that is a subset of the whole numbers. Justify your answers.

    a. How many students are in this class?

    b. How tall is each student?

    c. How old is each student?

    d. How many brothers does each student have? How many sisters?

    e. How many dimes are needed to purchase a can of pop from a vending machine in the cafeteria?

    f. How many buses are needed to transport 457 students?

16. Write a story problem whose answer is not a whole number.

17. Write the symbol for the empty set and write a practical example of an empty set.

Concept Map

Construct a concept map centered on the phrase **whole number**.

Reflection

Use your concept map to write a paragraph describing the important aspects of the set of whole numbers.

Section 2.2 Order of Operations with Whole Numbers

Purpose

- Investigate order of operations.
- Introduce some basic properties of whole numbers.
- Investigate computation on a calculator.

Investigation

1. John borrowed $12 from his brother. John spent $2 to buy two lottery tickets—the first ticket quadrupled his remaining money and the second added six dollars to his total. John then split the money he had with his brother, Mike. How much money did each man get?

   a. Indicate the computations you did to solve this problem.

   b. Write *one* expression that contains all the computations needed to solve this problem.

   c. Record the answer to the problem.

2. Jennifer had $12. She bought two paperback books that cost $4 each and cashed a check for six dollars. Jennifer combined the money she had with the money from the cashed check and divided it into two equal parts. She put one part in a secret compartment in her wallet. How much money did she hide?

   a. Indicate the computations you did to solve this problem.

   b. Write *one* expression that contains all the computations needed to solve this problem.

   c. Record the answer to the problem.

3. a. How are the problems in Investigations 1 and 2 the same?

   b. How are they different?

   c. Did your number expressions look the same or different?

4. Find the answer to $12 - 2 \times 4 + 6 \div 2$ without a calculator.

5. Enter $12 - 2 \times 4 + 6 \div 2$ on your calculator and record the answer.

6. Are the answers to Investigations 4 and 5 the same? If not, explain why.

7. Clearly describe the order of operations used by the calculator to compute $12 - 2 \times 4 + 6 \div 2$.

8. Evaluate each of the following and record the results in Table 1. In column 2, record the answer you found without the calculator. In column 3, record the answer returned by your calculator.

Problem	Answer without calculator	Answer with calculator
$13 - 5 - 2$		
$13 - (5 - 2)$		
$2 + 3(4)$		
$(2 + 3)4$		
$2 + (3(4))$		
$18 - 12 \div 3$		
$(18 - 12) \div 3$		
$\frac{20}{2 + 3}$		
$20 \div 2 + 3$		
$\frac{5 + 2}{8 - 1}$		
$\frac{3 \times 2^4}{11 - 3}$		
$4(7) + 2(20 - 7)$		
$2^7 - 3(11)$		

Table 1: Order of operations 1

9. For each problem in Investigation 8, record in Table 2 the operations and the order in which they are performed on the calculator.

Problem	Order of operations on the calculator
$13 - 5 - 2$	
$13 - (5 - 2)$	
$2 + 3(4)$	
$(2 + 3)4$	
$2 + (3(4))$	
$18 - 12 \div 3$	

Table 2: Order of operations 2

Problem	Order of operations on the calculator
$(18 - 12) \div 3$	
$\frac{20}{2 + 3}$	
$20 \div 2 + 3$	
$\frac{5 + 2}{8 - 1}$	
$\frac{3 \times 2^4}{11 - 3}$	
$4(7) + 2(20 - 7)$	
$2^7 - 3(11)$	

Table 2: Order of operations 2

10. Star each problem in Table 2 in which the parentheses used affects the output.

11. Identify the problems in Table 2 in which parentheses must be added as the problem is entered into the calculator to ensure correct output.

Discussion

The numbers used in Investigations 1 and 2 (page 60) were the same. Even the operations were the same, but were performed in different orders. In a situation, the action in the problem tells us what order in which to do our calculations. When we try to write the sequence of operations in a single number expression, we sometimes have difficulty.

In Investigation 1:

Subtract $2 from $12:	$12 - 2 = 10$
Quadruple the money:	$4 \times 10 = 40$
Add six dollars:	$40 + 6 = 46$
Split the money in half:	$46 \div 2 = 23$

In Investigation 2:

Multiply before subtracting:	$2 \times 4 = 8$
Subtract 8 from 12:	$12 - 8 = 4$
Add six dollars:	$4 + 6 = 10$
Split the money in half:	$10 \div 2 = 5$

The calculations in both investigations seem to be: $12 - 2 \times 4 + 6 \div 2$, the same sequence of numbers listed in Investigation 5, which, when entered on the calculator, displays the result 7. Looking at the single number expression $12 - 2 \times 4 + 6 \div 2$, a sequence of operations must be performed. The operations of addition, subtraction, multiplication, and division all occur in this problem and we get a different answer, depending on the order in which the operations are performed.

Mathematicians have agreed that the ***algebraic order of operations*** will be used. The agreed-upon order is: Multiply and divide from left to right before adding and subtracting, also from left to right. Most calculators use this order of operations as well.

In order to model what occurred in Investigations 1 and 2 mathematically, we must change the order of operations by using grouping (inclusion) symbols such as parentheses, brackets, or braces. We modify the order of operations by doing what is enclosed in parentheses first.

The number expression which models the problem in Investigation 1 should be written as:

$$[(12 - 2) \times 4 + 6] \div 2$$

and the number expression that models the problem in Investigation 2 needs to be written as:

$$[12 - 2 \times 4 + 6] \div 2$$

We can model what the calculator did in Investigation 5 by the use of inclusion symbols as well. Sometimes we insert parentheses to indicate how the calculations should be performed, even though the results would be the same without the use of inclusion symbols. In investigation 5, the computations could be described as follows:

$$12 - (2 \times 4) + (6 \div 2)$$

The operations of addition, subtraction, multiplication, and division are ***binary*** operations, which require two ***inputs*** to produce one ***output***. For example, in the computation $3 + 5 = 8$, the operation of addition is used on the inputs 3 and 5 to produce the output 8. When more than one operation is indicated, we first choose to do one operation with its two inputs, produce an output, then do the next operation. For example, in the computation $4 + 3 - 2$, we add $4 + 3$ to get the output 7, then subtract $7 - 2$ to get our final output, 5.

The computations in the preceding investigations involve two basic operations. However, by performing operations from left to right, we may not always obtain the generally agreed–upon answer. Since we get a different answer depending on the *order* in which the operations are performed in many of these problems, it is important we all agree on the order of operations which proceeds from the most complex operations to the most basic operations. The algebraic order of operations will be used throughout the course.

Table 3 gives answers to the preceding investigations. Please note that there are several equivalent forms of statements for each problem though only one English statement is listed for each problem. The asterisk indicates a problem in which the parentheses affect the output.

Problem	Answer	Order of Operations	English statement
$13 - 5 - 2$	6	Subtraction followed by subtraction.	Subtract 5 from 13 and subtract 2 to the difference.
* $13 - (5 - 2)$	10	Subtraction followed by subtraction.	Subtract the difference of 5 and 2 from 13.
$2 + 3(4)$	14	Multiplication followed by addition.	Add 2 to the product of 3 and 4.
* $(2 + 3)4$	20	Addition followed by multiplication.	The product of 4 and the sum of 2 and 3.
* $(18 - 12) \div 3$	2	Subtraction followed by division.	Subtract 12 from 18. Divide the difference by 3.
$18 - 12 \div 3$	14	Division followed by subtraction.	Divide 12 by 3. Subtract the quotient from 18.
$\frac{20}{2 + 3}$	4	Addition followed by division.	20 divided by the sum of 2 and 3.

Table 3: Describing computations

Problem	Answer	Order of Operations	English statement
$20 \div 2 + 3$	13	Division followed by addition.	Divide 20 by 2. Add 3 to the quotient.
$\frac{5+2}{8-1}$	1	Addition, subtraction, and division	Divide the sum of 5 and 2 by the difference of 8 and 1
$\frac{3 \times 2^4}{11-3}$	6	Exponentiation, multiplication, subtraction, division.	Raise 2 to the fourth power. Multiply the result by 3. Divide the product by the difference of 11 and 3.
* $4(7) + 2(20-7)$	54	Subtraction, multiplication, multiplication, addition.	Subtract 7 from 20. Multiply 7 by 4. Multiply the difference of 20 and 7 by 2. Add the two products.
$2^7 - 3(11)$	95	Exponentiation, multiplication, subtraction.	Raise 2 to the seventh power. Subtract the product of 3 and 11 from this answer.

Table 3: Describing computations

Note the last two rows of Table 3. The expression $4(7) + 2(20-7)$ was used in studying the parking lot problem in Section 1.1. This is an expression for the total number of wheels in the parking lot if there are seven cars and thirteen motorcycles.

The expression $2^7 - 3(11)$ was used in Section 1.4 to represent the P–count of a S P C valid path if Rule 1 had been applied seven times and Rule 3 had been applied eleven times beginning with the initial path S P.

The order in which you enter numbers into a symbolic manipulator or scientific calculator may require you to enter parentheses that are not indicated in the problem. For example, in $\frac{5+2}{8-1}$, the fraction bar acts like a ***symbol of inclusion.*** Notice that the fraction bar separates the numerator, $5+2$, from the denominator, $8 + 1$. Add $5+2$ getting a single value, 7, as the new numerator. Likewise, subtract $8 - 1$ to get a single value, 7, in the denominator, before doing the division. Entered in the calculator or computer without parentheses, the problem reads $5 + 2 \div 8 - 1$ and results in the answer 4.25. Parentheses must be added when entering the problems $\frac{20}{2+3}$, $\frac{5+2}{8-1}$, and $\frac{3 \times 2^4}{11-3}$ into the calculator. The parentheses are added to group the denominators in all of these problems. Parentheses must also be added to the numerator in $\frac{5+2}{8-1}$ to group the sum of 5 and 2.

Parentheses are one way to change the algebraic order of operations. When the calculations include exponentiation, radicals, or fractions, by agreement, operations will be performed as follows.

Order of operations

1.	Operations inside grouping symbols. The grouping symbols include both parentheses and fraction or division bars. If there are several operations within a grouping symbol, the operations are done in the order specified in steps 2–4.
2.	Exponentiation from left to right.
3.	Multiplication and division from left to right.
4.	Addition and subtraction from left to right.

Ways to change the order of operations are explored in the following investigations.

Investigation

12. Record the order in which you do the following computations mentally.

    a. $6 + 7 + 8 + 4 + 2 + 3$

    b. $4 \times 17 \times 25$

13. Sue finds the answer to $9 + 6$ by computing $10 + 5$.

    a. Explain why Sue's method works.

    b. Use Sue's method to find the answer to $9 + 8$. Write the problem you actually did.

    c. Use Sue's method to find the answer to $99 + 57$. Write the problem you actually did.

14. Write the answers to each problem.

a. $12 + 3$

$3 + 12$

b. $12 - 3$

$3 - 12$

c. $12\,(3)$

$3\,(12)$

d. $12 \div 3$

$3 \div 12$

15. Refer to the examples in Investigation 14 to answer the following.

a. How are the two expressions in each part the same? How are they different?

b. For which of the four operations does the order of the inputs not affect the output?

c. For which of the four operations does the order of the inputs affect the output?

16. Write the answers to each problem.

a. $(32 + 8) + 4$

$32 + (8 + 4)$

b. $(32 - 8) - 4$

$32 - (8 - 4)$

c. $(32\,(8))\,4$

$32\,(8\,(4))$

d. $(32 \div 8) \div 4$

$32 \div (8 \div 4)$

17. Refer to the examples in Investigation 16 to answer the following.

a. How are the two expressions in each part the same? How are they different?

b. For which of the four operations does the grouping of three inputs not affect the output?

c. For which of the four operations does the grouping of three inputs affect the output?

Discussion

In each of the above Investigations, you discovered, or have been taught, a legal way of breaking the rules which gives you a correct answer. For example, in Investigation 10 you probably grouped numbers to make sums of 10, thus changing the order. The problem is much easier if thought of as

$$(6+4) + (7+3) + (8+2)$$

The order of computation in the original problem is changed, as well as the way the numbers are grouped. Instead of adding from left to right, group terms which sum to 10, then add up the three sums to get the answer 30. When you *change the order* in which you add you are using the ***commutative property of addition*** and when you *change the grouping* of numbers, you are using the ***associative property of addition***, which does not change the order in which the numbers are listed. In both instances, the sum is not affected.
Sue thought of 9 + 6 as the problem $9 + (1+5)$ allowing her to mentally regroup the numbers. Instead of summing the 1 with 5, 1 is paired with 9, using the associative property of addition, i.e., $(9+1)+5$.

We could also change $4 \times 17 \times 25$ to $17 \times 4 \times 25$ since the order of the ***factors*** does not affect the product. Once we change the order of the numbers using the ***commutative property of multiplication,*** we mentally regroup $17 \times (4 \times 25)$ for easier multiplication, using the ***associative property of multiplication.*** The word "commute" implies the reversal of order, "associate" implies grouping.

These investigations provided numerical examples of some important properties of the whole number system, namely the associative properties of addition and of multiplication, as well as the commutative properties of addition and multiplication. Note that the operations of subtraction and division of whole numbers are not associative nor commutative.

Investigation

18. Write the answers to each problem.

a. $2(3+5)$

$2(3)+5$

$2(3)+2(5)$

b. $4(7-2)$

$4(7)-2$

$4(7)-4(2)$

19. In Investigation 18, which two problems in each part are ***equivalent***?

20. Describe an easy way to break up 9 so that the following multiplication can be done mentally, without paper and pencil or a calculator.

    367×9

Discussion

We can do many problems mentally, if we change the problem to an equivalent problem, using the order of operations. The computation 367×9 is the same as $367 \times (10 - 1)$ or $3670 - 367$. In fact we used the ***distributive property*** to do this computation. This property not only forms the basis of our algorithm for multiplication, but is one of the most important ways of breaking the rules for the order of operations. In the 1500's, the mark of an educated person was the ability to memorize all of the multiplication facts from 0 x 0 through 99 x 99. The distributive property showed how two digit multiplication could be done taking one digit times one digit. This discovery meant that the task of memorizing 8100 multiplication facts could be replaced by a single process using the distributive property.

In Investigation 18, note that

$$2(3+5) = 2(3) + 2(5)$$

and

$$4(7-2) = 4(7) - 4(2).$$

These are examples of, respectively, the

distributive property of multiplication over addition

and

distributive property of multiplication over subtraction.

The word "distribute" implies that the factor outside the grouping symbol is distributed over each input to the sum or difference operation. We note that multiplication does not distribute over multiplication or division.
There are several issues to be aware of when using a calculator to perform a computation.

- Does your calculator round decimal values or truncate them to fit the display window?
- How many digits does your calculator display?
- How does your calculator indicate an illegal operation was performed?

Investigation

21. Enter $2 \div 3$ on your calculator. Record the displayed answer and the number of digits. What is the last digit?

22. Multiply your answer to Investigation 21 by 3. Record your answer and the last digit.

23. Enter $8 \div 0$ on your calculator. Record your answer.

Discussion

If your calculator displayed a 7 as the last digit when $2 \div 3$ was entered, your calculator rounds off the answer. If your calculator displayed an answer containing only the digit 6 repeated, your calculator ***truncates*** the answer.

Likewise, if you multiply the answer to the division $2 \div 3$ by 3 and the last digit is a 9, or the displayed answer is a 2, your calculator rounds off the answer. If the last digit is 8, your calculator truncates the answer.

Your calculator should return an error message when $8 \div 0$ is entered since division by 0 is an undefined operation. Note how the error message is indicated on your calculator.

Explorations

1. List and define words in this section that appear in ***italics bold*** type using your own words.
2. Indicate which definitions you knew and which definitions you found in an outside source. Indicate your source.
3. Perform the following calculations.

a. $5(7+6)$

b. $18-2(3+4)$

c. $9-3(2)$

d. $\dfrac{8}{9-5}$

e. $28+10 \div 5$

f. $17-3+5+8-2$

g. $11(5)-3(4)$

h. $(8+10) \div 4+3(2)$

i. $\dfrac{10+8}{5+4}$

4. What happens when we use exponents or take roots in a problem which has parentheses or which involves fractions? Evaluate the following:

$$\frac{5 + 2\,(4^3 - 6)}{\sqrt{8+1} + 7} - 2$$

5. Use a commutative property to rewrite the following.

   a. $18 + 9 =$ b. $7\,(11) =$

6. Use an associative property to rewrite the following.

   a. $5 + (3 + 7) =$ b. $(5\,(3))\,7 =$

7. Use a distributive property to rewrite the following.

   a. $7\,(4 + 9) =$ b. $4\,(8 - 3) =$

   c. $5\,(4) + 5\,(11) =$ d. $16\,(3) - 16\,(2) =$

8. Write your own example of each of the following.

   a. The associative property of addition.

   b. The commutative property of multiplication.

   c. The distributive property of multiplication over subtraction.

9. Does the operation of division distribute over addition? over subtraction? Create some numeric examples as data to help answer this question.

10. Let the variable n represent any whole number. Write an algebraic expression for:

    a. three more than n.

    b. five times n.

    c. eight less than n.

    d. two more than the product of 7 and n.

11. Refer to the expressions you wrote as answers to Exploration 10. Record the value of each of the expressions if n has a value of six.

12. Use the calculator keys 1, 7, and 9 exactly once, in any order, with any

12. Use the calculator keys 1, 7, and 9 exactly once, in any order, with any operation allowable on the whole numbers, to find as many counting numbers 1–20 as possible.

13. Compare the answer to $\frac{20}{2 \cdot 3}$ with the answer to $20/2 \cdot 3$. Explain what is happening in terms of order of operations.

14. Enter a counting number less than 20. If the number is even, divide it by 2; if odd, multiply it by 3 and add 1. Use the answer displayed on your calculator to repeat the process.

Here's an example: Suppose you choose 14. Since it is even, divide by 2 to get 7. Seven is odd so multiply it by 3 (21) and add 1 to get 22. Twenty–two is even so divide it by 2 to get 11. This is odd so multiply 11 by 3 and add 1 to get 34. And so on.

a. Can you reach the number 1? Explain.

b. Which number requires the most steps? How many steps were required?

c. Analyze the results using an even number and compare them with the results using an odd number. What do you observe?

d. Repeat the process using a number larger than 20. What happens?

Concept Map

Construct a concept map centered on the phrase **operations on whole numbers**.

Reflection

Use your concept map to write a paragraph describing the order of operations and the properties of the operations on the whole numbers.

Section 2.3 Algebraic Extensions of the Whole Numbers

Purpose

- Introduce algebraic expressions and their meaning.
- Evaluate algebraic expressions.
- Generalize the commutative, associative, and distributive properties.

Investigation

1. a. Store the whole number 3 in your calculator as x.

   b. ***Evaluate*** the expression $7x$ and record the answer.

   c. Choose a whole number and store it in your calculator as x. Evaluate $7x$. Record the number you chose and the answer.

   d. What operation is being performed on x?

   e. Describe in a complete sentence the meaning of $7x$.

   f. Draw a picture that models $7x$.

2. a. Store the whole number 3 in your calculator as x.

   b. Evaluate the expression $5 + x$ and record the answer.

   c. Choose a whole number and store it in your calculator as x. Evaluate $5 + x$. Record the number you chose and the answer.

   d. What operation is being performed on x?

e. Describe in a complete sentence the meaning of $5 + x$.

f. Draw a picture which models $5 + x$.

3. a. Store the whole number 3 in your calculator as x.

   b. Evaluate the expression $5 + 7x$ and record the answer.

   c. Choose a whole number and store it in your calculator as x. Evaluate $5 + 7x$. Record the number you chose and the answer.

   d. What operations are being performed on x and in what order?

   e. Describe in a complete sentence the meaning of $5 + 7x$.

4. a. Store the whole number 3 in your calculator as x.

   b. Evaluate the expression $4x^2$ and record the answer.

   c. Choose a whole number and store it in your calculator as x. Evaluate $4x^2$. Record the number you chose and the answer.

   d. What operations are being performed on x and in what order?

   e. Describe in a complete sentence the meaning of $4x^2$.

5. a. Store the whole number 3 in your calculator as x.

   b. Evaluate the expression $3 + 5x^2 - 2x$ and record the answer.

   c. Choose a whole number and store it in your calculator as x. Evaluate $3 + 5x^2 - 2x$. Record the number you chose and the answer.

   d. What operations are being performed on x and in what order?

   e. Describe in a complete sentence the meaning of $3 + 5x^2 - 2x$.

Discussion

Before we discuss the investigations, note that we are now performing operations on expressions containing variables. This allows us to make some generalizations symbolically.

For example, if m and n represent whole numbers, we represent their sum by $m + n$. In this expression, m and n are called ***terms***. We represent the difference between m and n by the expression $m - n$.

Both the sum $m + n$ and the difference $m - n$ can be thought of two different ways:

- $m + n$ can be thought of as two terms on which the binary operation of addition (**process**) is performed, or $m + n$ can be considered as an **answer**, the sum $m + n$.

- Similarly, $m - n$ can be thought of as the **process** of subtracting n from m or as an **answer**, the difference $m - n$.

The product of the whole numbers m and n is written several ways, the most common of which are mn, $m(n)$, $(m)n$, or $(m)(n)$. Regardless of the notation, m and n are called the ***factors*** of this expression.

The quotient of the whole numbers m and n is written as $\frac{m}{n}$ or $m \div n$.

The four basic binary operations, each of which receives two inputs and has one output, are demonstrated by a visual aid called a ***function machine*** (Figure 1).

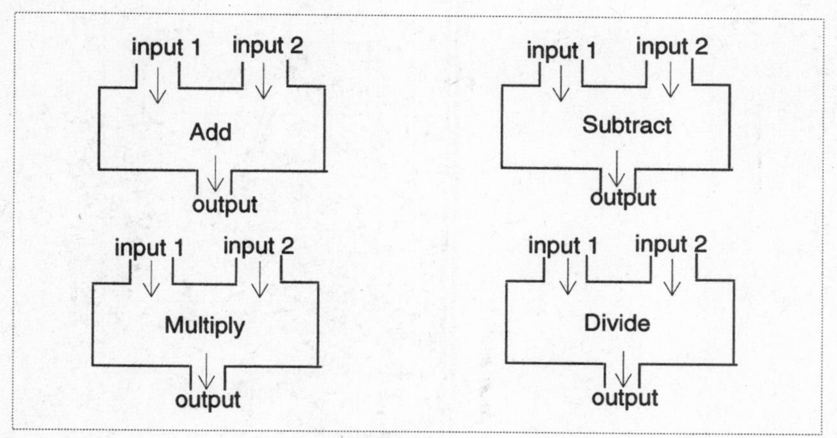

Figure 1

In the preceding investigations, the output $7x$ represents the product of 7 and the unknown number x. We could also have written $7(x)$, $(7)x$, or $(7)(x)$. Given a value for x, we get a value for $7x$ (Figure 2).

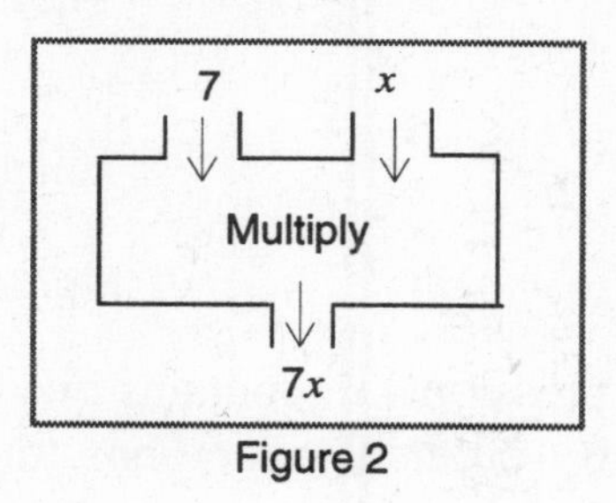

Figure 2

The expression $5 + x$ represents the sum of 5 and the unknown number x (Figure 3).

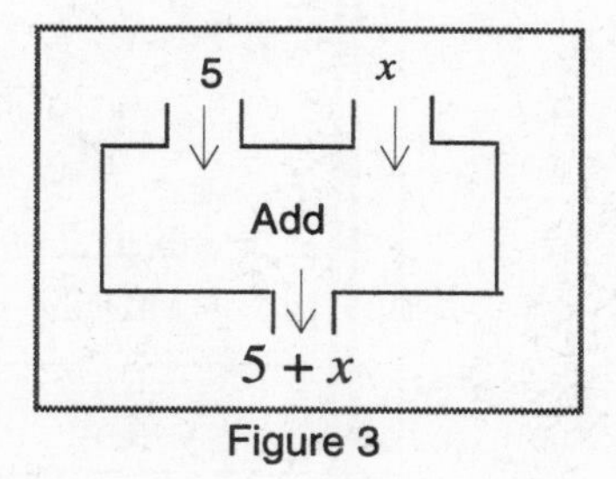

Figure 3

The expression $5 + 7x$ contains two operations, addition and multiplication. By order of operations, we multiply the unknown number x by 7 first, then add 5 to the product. So, $5 + 7x$ can be read as the sum of 5 and the product of 7 times x. (Figure 4).

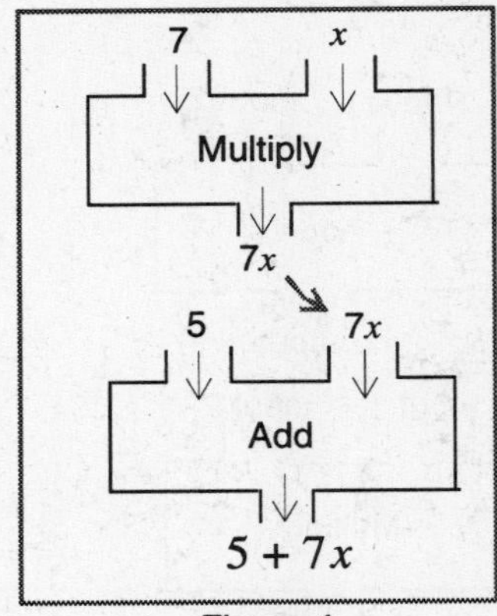

Figure 4

Some operations require a single input with one output. Such operations are called ***unary*** operations. Exponentiation is considered a unary operation and is represented by the function machine in Figure 5.

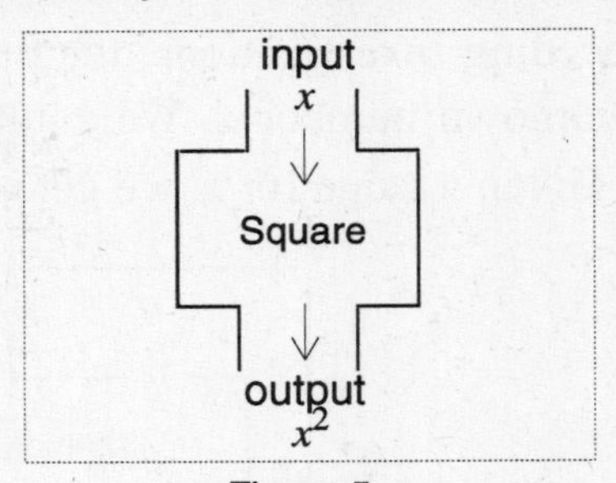

Figure 5

The expression $4x^2$ contains two operations, multiplication and exponentiation. By order of operations, the exponentiation is performed first, followed by the multiplication. So $4x^2$ represents the product of 4 and the square of the unknown number x (Figure 6).

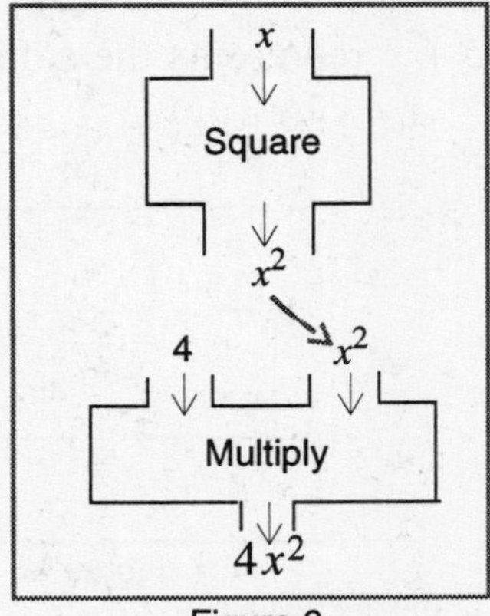

Figure 6

The expression $3 + 5x^2 - 2x$ contains the operations of addition, multiplication, subtraction, and exponentiation. The order of operations convention requires that the value of x first be squared, then multiplied by 5 to get $5x^2$. The input x must also be multiplied by 2 to get $2x$. The ***constant*** 3 is added to $5x^2$ to get $3 + 5x^2$. Finally $2x$ is subtracted from $3 + 5x^2$. Wow, that's a lot of steps! Figure 7 contains the function machine for this expression.

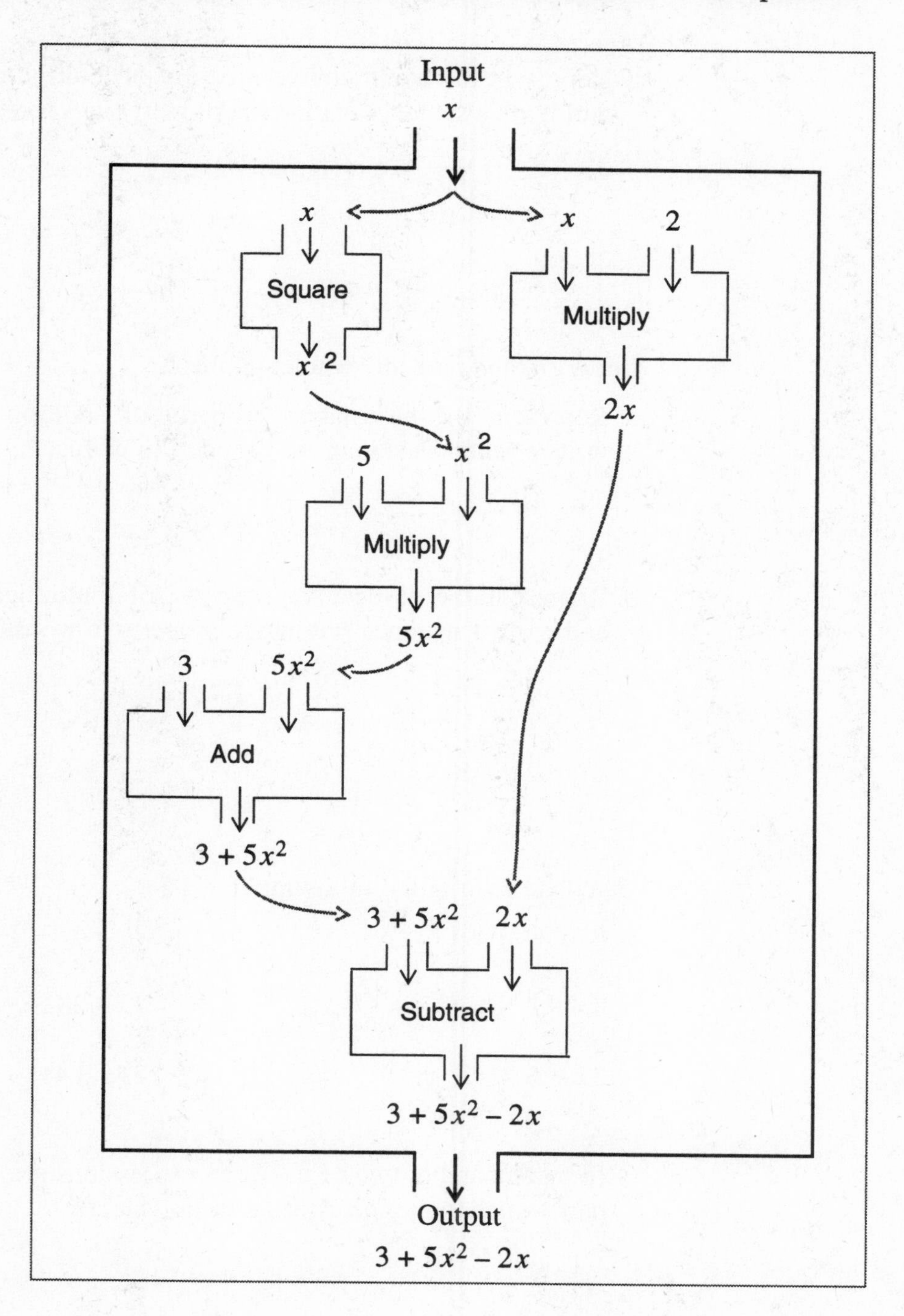

Figure 7

Investigation

6. Let m and n be any whole numbers.

   a. Represent the commutative property of addition symbolically and write a sentence stating the property in words.

   b. Represent the commutative property of multiplication symbolically and write a sentence stating the property in words.

7. Let m , n, and p be any whole numbers.

   a. Represent the associative property of addition symbolically and write a sentence stating the property in words.

   b. Represent the associative property of multiplication symbolically and write a sentence stating the property in words.

8. Calculate each of the following if x is 5.

a. $3(x+2)$	b. $7x-4$
$3x+2$	$7x-28$
$3x+6$	$7(x-4)$

   c. In parts a and b, two of the three expressions are equivalent. In each part. which two expressions are equivalent?

9. Let m, n, and p represent any whole numbers.

   a. Write the distributive property of multiplication over addition symbolically.

   b. Write a sentence stating the distributive property of multiplication over addition in words.

Discussion

The power of algebra lies in its ability to generalize. In the mathematical equation $2 + 3 = 3 + 2$ we know the sum of the numbers 2 and 3 is 5, regardless of the order of the ***addends***. However, using variables, we can state an equivalent principle for any pair of whole numbers. Similarly, we can write a generalized statement for each of the other properties we have investigated. Let m, n, and p represent any whole numbers. Then the properties are:

- the commutative property of addition: $m + n = n + m$
- the commutative property of multiplication: $mn = nm$
- the associative property of addition: $(m + n) + p = m + (n + p)$
- the associative property of multiplication: $(mn)p = m(np)$
- the distributive property of multiplication over addition: $p(m + n) = pm + pn$
- the distributive property of multiplication over subtraction: $p(m - n) = pm - pn$.

Explorations

1. List and define words in this section that appear in ***italics bold*** type using your own words.

2. Evaluate each of the following if $p = 5$

   a. $2p$
   b. p^2
   c. $p + 11$
   d. $6 - p$
   e. $3 + 2p$
   f. $4p - 7$
   g. $p^2 + 3$
   h. $3p^2 - 4$
   i. $p^2 + 6p$
   j. $6 + 7p + p^2$
   k. $9p - 2p^2 + 8$

3. For parts e, h, and j in Exploration 2, do the following:

   a. List the operations and the order in which they are performed.
   b. Describe the computation in words.
   c. Draw a function machine representation.

4. Rewrite using the distributive property of multiplication over either addition or subtraction.

   a. $7\,(x + 2)$
   b. $5\,(8 - m)$
   c. $6b - 18$
   d. $35 + 7t$

5. Given the function machine, evaluate when the input is 7.

   a.

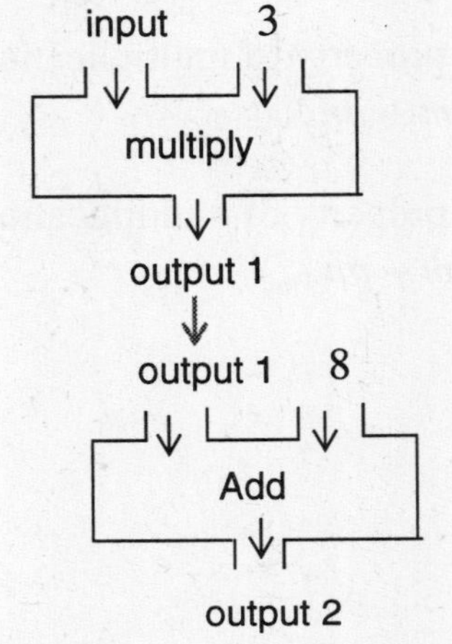

b.

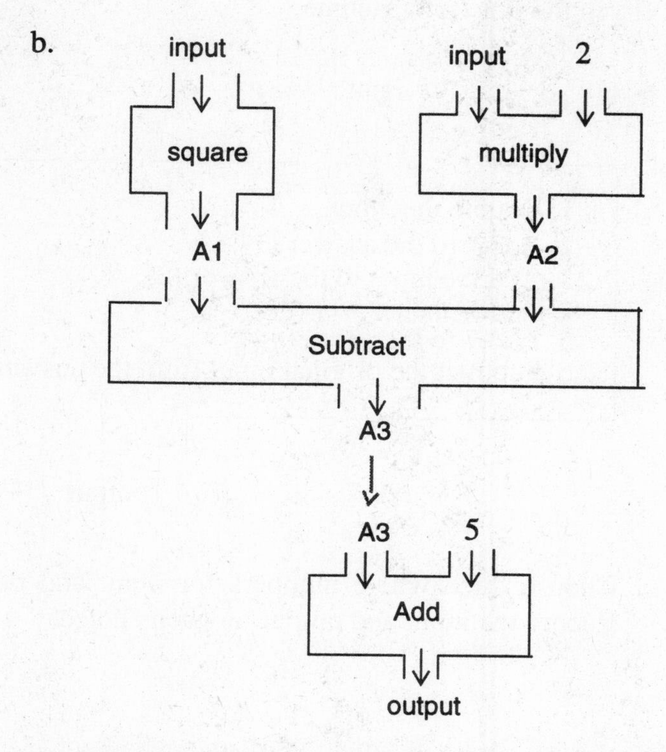

6. For each function machine in Exploration 5, let the variable x represent the input. Write an algebraic expression that represents the function machine.

7. Draw a function machine for each of the rules in the S P C system. For each give an example of a theorem that could serve as input and show the corresponding output.

8. Create a function machine for the following:

   a. The number of wheels in the full twenty space parking lot where the number of wheels equals $4c + 2(20 - c)$. The variable c represents the number of cars in the lot.

   b. The P–count of a S P C valid path as given by the expression $2^n - 3x$ where n represents the number of times Rule 1 is applied and x represents the number of times Rule 3 is applied. Be careful—this expression has two inputs.

9. Given the function machine.

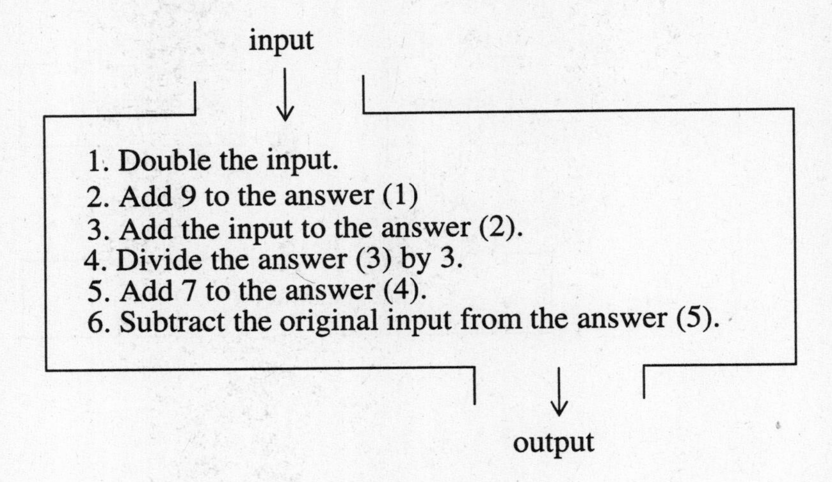

a. Choose three whole numbers for input and determine the output. Record the input and output for each choice.

b. Let the input variable be n. Write an algebraic expression for the function machine.

Concept Map

Construct a concept map centered on the phrase **algebraic expression.**

Reflection

Create your very own algebraic expression that involves the operations of addition, subtraction, multiplication, and exponentiation and contains one variable. Describe the expression in words explaining how you would evaluate it for a given value of the variable. Draw a function machine for your expression.

Section 2.4 Making Connections: What does it mean to Generalize?

Purpose

- Reflect upon ideas explored in Chapter 2.
- Explore the connection between doing mathematics and generalizing.

Investigation

In this section you will work outside the system to reflect upon the mathematics in Chapter 2: what you've done and how you've done it.

1. State the five most important ideas in this chapter. Why did you select each?

2. Identify all the mathematical concepts, processes and skills you used to investigate the problems in Chapter 2?

3. What has been common to all of the investigations which you have completed?

4. Select a key idea from this chapter. Write a paragraph explaining it to a confused best friend.

5. You have investigated many problems in this chapter.

   a. List your three favorite problems and tell why you selected each of them.

   b. Which problem did you think was the most difficult and why?

6. What does it mean to generalize? How do you test a generalization?

Discussion

There are a number of really important ideas which you might have listed including whole number, order of operations, properties of whole numbers, function machines and algebraic expressions.

Concept Map

Construct a concept map centered around the phrase **generalization**.

Reflection

Find a computational error made in some problem you worked in Chapter 2. Discuss how over-generalization and/or an incorrect application of the order of operations contributed to your error.

Illustration

Draw a picture of **a whimsical mathematical creature**.

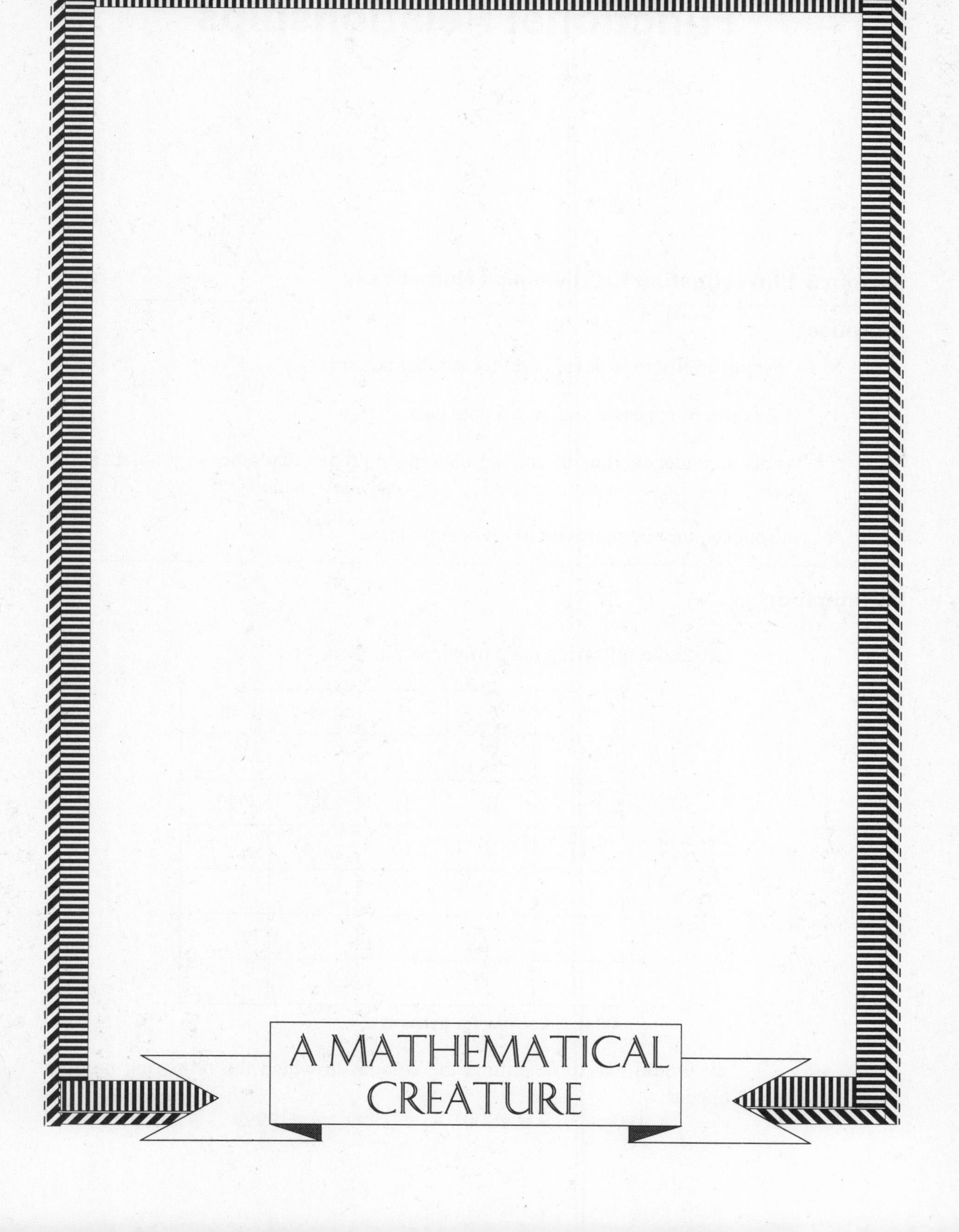
A MATHEMATICAL
CREATURE

Chapter 3

Functional Relationships

Section 3.1 Investigating Relationships Numerically

Purpose

- Develop ability to look for and find number patterns.
- Use tables to organize and investigate data.
- Apply an understanding of number patterns to predict outcomes in problem situations.
- Introduce numeric representations of functions.

Investigation

Given the following table of whole numbers

	2	3	4	5
9	8	7	6	
	10	11	12	13
17	16	15	14	

Table 1: Looking for patterns

we would like to determine the column in which the following numbers appear:

a. 100 b. 1000 c. 1999 d. 9999 e. 99997

1. Begin your investigation by studying the table. Describe any patterns you notice.

2. Complete another three rows of the table.

3. In what column does the number 36 appear? How did you determine this?

4. Would your method be a reasonable way to find the number 99,997? Explain.

5. Describe one characteristic all numbers in column 1 have in common.

6. Describe any patterns you observe about the numbers in another column.

7. There are five columns in this array. Are there five different patterns? If not, how many different patterns are there?

8. Based on your observations, can you predict in which columns the numbers 100 and 1000 will **not** occur? How about the numbers 1999, 9999, and 99997?

Discussion

This is a problem for which an equation is not obvious. There are numerous ways to investigate a problem. If a rule is given, the problem may be represented using mathematical statements. Most often, the rule is not given and it must be determined using data. A way to begin the process is to look for patterns in the data. Rules may be generalized from the patterns. When mathematics is practiced, it often originates from the generalization of patterns recognized in data.

Look for a pattern.

Problem Solving Toolkit

Looking at columns of numbers to determine if they contain a collection of odd numbers, or a collection of even numbers, is one way to investigate relationships. Studying the table reveals that we have two simpler problems to solve.

- Recognizing that even numbers only occur in columns 2 and 4 results in the simpler problem: does 100 occur in column 2 or 4?
- Since odd numbers occur only in columns 1, 3, or 5, we only need to determine in which of the three odd columns the numbers 1999, 9999, and 19,997 occur.

Reduce a complex problem to a simpler problem.

Problem Solving Toolkit

Investigation

9. In the first column of Figure 1, subtract 9 from 17 and then subtract 17 from 25. Record your answers on the blank lines provided. Repeat this process of subtraction in each column, subtracting the first number in the column from the second, the second number from the third, etc.

Record each ***difference*** to the right on the line provided.

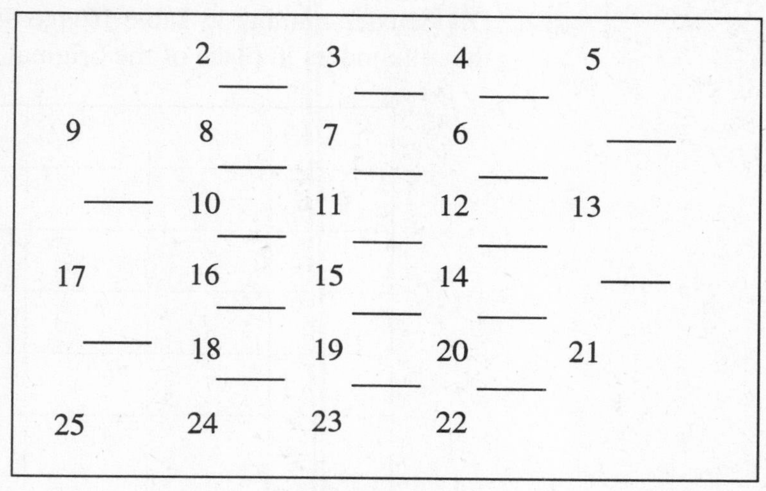

Figure 1

10. Look at the differences. Does the same difference occur in more than one new column? If so, does the difference occur in the same way?

11. What number or numbers appear important in the investigation? Justify your choice.

Discussion

The process of subtracting a preceding number in a column from the next number in a column is called a process of ***finite differences***. This process is used to discover relationships between numbers. In the above investigation, the number 8 appears important. However it does not appear to have a connection with every number in the table. In the second and fourth columns, the numbers 2 and 6 appear as alternating finite differences. Yet, $2 + 6 = 8$, so there seems to be a connection between the number 8 and the differences in each column.

So far we have used subtraction to find finite differences between the numbers in the columns and we have used addition of some finite differences to find a connection between the number 8 and finite differences in the other columns.

Investigation

12. Divide each number in Table 1(page 88) by 8. In Table 2, record only the remainders in place of the original numbers.

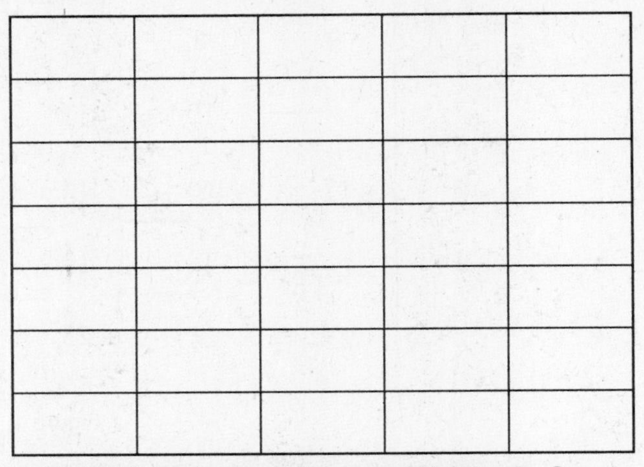

Table 2: Table of remainders upon dividing by 8

13. Using Table 2, describe in writing how you can determine the column in which a given number will appear.

14. Using Table 2, predict in which columns the numbers 100, 1000, 1999, 9999, and 99997 will occur.

Discussion

In the preceding investigation, we used finite differences, addition, and division to search for a pattern. Recognition and generalization of the pattern allowed us to answer what originally appeared to be a difficult problem. Mathematics can be thought of as the science of patterns and order of all sorts. Algebra is a tool we can use to describe and analyze patterns discovered when we investigate relationships. It is the key to understanding mathematical structures.

The pattern allows us to be given a whole number and specifically apply a rule to obtain the column in which the number appears. This provides us with a relationship between two changing quantities: a whole number greater than one and the column in which the whole number appears in the table. Furthermore, a choice of a whole number determines exactly one column number. Such a relationship is called a ***function***.

A function is a process that receives ***input*** (a whole number greater than one in this case) and returns a unique value called the ***output*** (a column number in this case). The set of possible inputs to the function is called the ***domain*** of the function. The set of outputs of the function is called the ***range*** of the function. In this chapter we explore functions using tables, algebraic statements involving variables, and graphs. We also regularly use function machines to illustrate these relationships. For example, Figure 2 displays a general function machine for this problem.

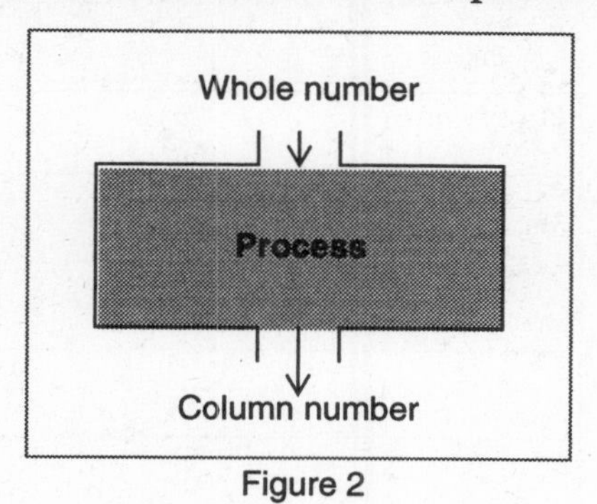

Figure 2

The next set of investigations will lead you through the process of clearly defining the process for the function machine displayed in Figure 2.

Investigation

Input/output tables are often used to numerically represent relationships between two changing quantities. Let's create several tables that can be used to investigate the problem in this section.

15. Complete Table 3.

Whole number	Remainder when whole number is divided by 8
57	
934	
1723	
4332	
19805	

Table 3: Remainder as determined by whole number

16. Construct a function machine with input a whole number and output the remainder of the whole number when the whole number is divided by 8.

17. Complete Table 4. Record the remainders from Table 3 in the input (left) column. In the output (right) column, record the column from Table 1 that corresponds to the remainder.

Remainder when whole number is divided by 8	Column in which whole number appears

Table 4: Column as determined by remainder

18. Construct a function machine with input a remainder when a whole number is divided by 8 and output the column in which the whole number would appear in Table 1 (page 88).

19. Complete Table 5 by combining Tables 3 and 4.

Whole number	Column in which whole number appears
57	
934	
1723	
4332	
19805	

Table 5: Column as determined by whole number

20. Construct a function machine with input a whole number and output the column in which the whole number appears in Table 1 (page 88).

Discussion

Table 6 is a completed version of Table 3 (page 93).

Whole number	Remainder when whole number is divided by 8
57	1
934	6
1723	3
4332	4
19805	5

Table 6: Remainder as determined by whole number

A function machine for the process of using a whole number as input and receiving its remainder when the whole number is divided by 8 appears in Figure 3.

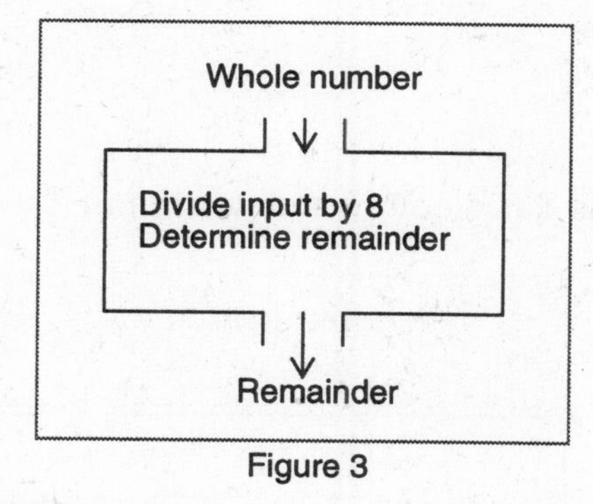

Figure 3

Table 7 is a completed version of Table 4 (page 94).

Remainder when whole number is divided by 8	Column in which whole number appears
1	1
6	4
3	3
4	4
5	5

Table 7: Column as determined by remainder

A function machine for this process of using a remainder from division by 8 to determine the column a number would appear in Table 1 (page 88) appears in Figure 4.

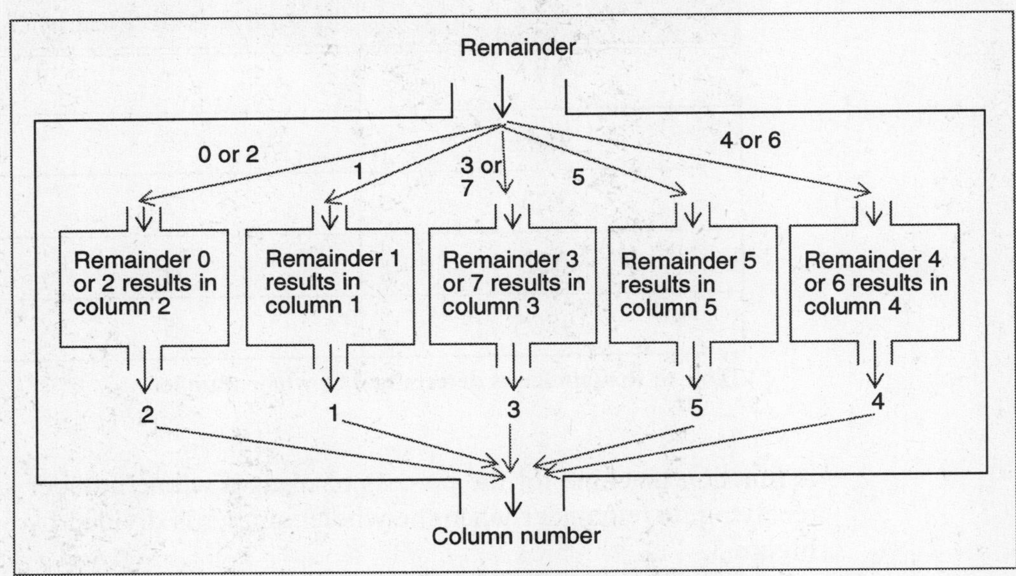

Figure 4

Table 8 is a completed version of Table 5 (page 94).

Whole number	Column in which whole number appears
57	1
934	4
1723	3
4332	4
19805	5

Table 8: Column as determined by whole number

A function machine for the process of using a whole number as input and determining the column where the whole number will occur in Table 1 (page 88) appears in Figure 5.

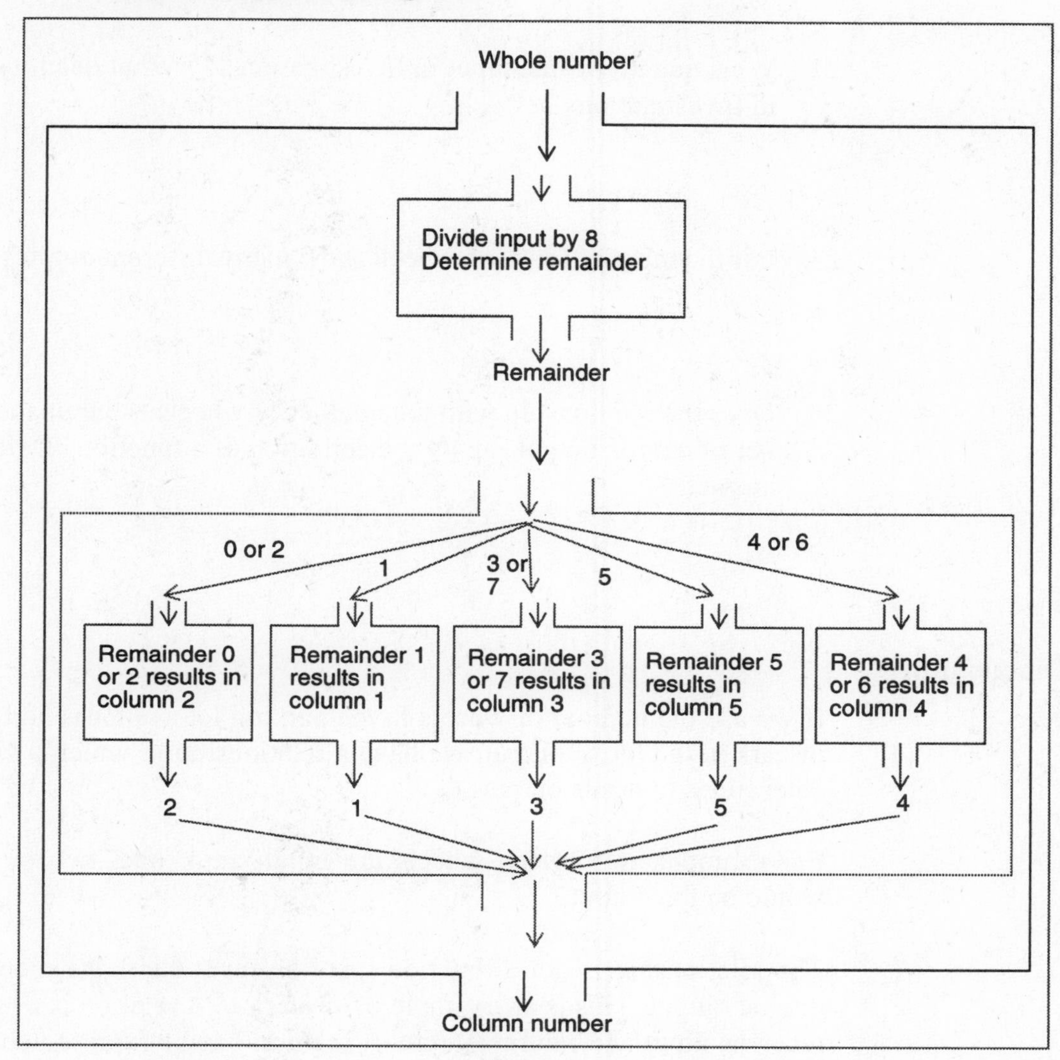

Figure 5

When function was defined on page 92, we stated that every input must have an *unique* output. Let's look at a relationship where this is not the case.

Investigation

There is a parking lot on campus with space for twenty vehicles which might be cars or motorcycles. The parking lot is not necessarily full. Let's investigate the relationship between the number of wheels in the lot and the number of cars in the lot.

21. If there are fourteen wheels in the lot, how many cars are in the lot?

22. What quantity is the input in Investigation 21? What quantity is output in Investigations 21?

23. For the input of fourteen wheels, how many different outputs are there?

24. Does the relationship with the number of wheels as input and the number of cars as output satisfy the definition of a function? Defend you answer.

Discussion

If we use the number of wheels in the parking lot as input and the number of cars in the lot as output, we have a relationship in which a single input determines multiple outputs.

For example, if fourteen wheels are input, zero, one, two or three cars would be the output.

The relationship is not a function since an input can have more than one unique output. This is an example of a ***relation***. A relation is a process that receives input and returns output. A relation can be represented by a relation machine in which multiple outputs might occur.

Figure 6 contains a relation machine for the situation in which fourteen wheels are input.

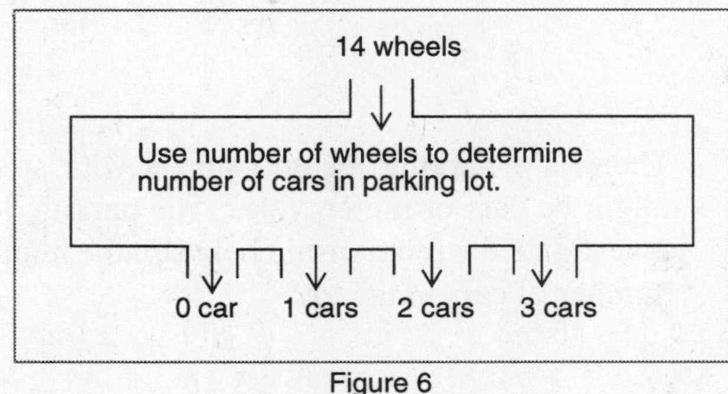

Figure 6

Explorations

1. List and define words in this section that appear in ***italics bold*** type using your own words.

2. Write the set that represents the domain of the function displayed in Figure 3 (page 95). Is the domain finite or infinite? Why?

3. Write the set that represents the range of the function displayed in Figure 3 (page 95). Is the range finite or infinite? Why?

4. Write the set that represents the domain of the function displayed in Figure 4 (page 96). Is the domain finite or infinite? Why?

5. Write the set that represents the range of the function displayed in Figure 4 (page 96). Is the range finite or infinite? Why?

6. Write the set that represents the domain of the function displayed in Figure 5 (page 97). Is the domain finite or infinite? Why?

7. Write the set that represents the range of the function displayed in Figure 5 (page 97). Is the range finite or infinite? Why?

8. The discussion on page 92 states that the domain for the problem situation is the set of whole numbers greater than one. Which whole numbers are excluded from this set?

9. Compare and contrast the definition of function (page 92) and the definition of relation (page 98).

10. Given the sequence of sums

$$1 = 1$$

$$1 + 3 = ?$$

$$1 + 3 + 5 = ?$$

$$1 + 3 + 5 + 7 = ?$$

   a. Write the next four rows in the sequence.

   b. Describe in a complete sentence the pattern you discover.

   c. Construct a table of five input/output pairs in which the input is the indicated sum of all odd numbers less than a given odd number (such as $1 + 3 + 5 + 7$) and the output is the number of odd numbers in the indicated sum.

   d. Draw a function machine with input the indicated sum of all odd numbers less than a given odd number (such as $1 + 3 + 5 + 7$) and output the number of odd numbers in the indicated sum.

e. Construct a table of five input/output pairs in which the input is the number of terms in the sum of the odd numbers and the output is the sum of the odd numbers.

f. Draw a function machine with input the number of terms in the sum of the odd numbers and output the sum of the odd numbers.

g. Use the function machines from parts d. and f. to draw a function machine with input the indicated sum of all odd numbers less than a given odd number (such as $1 + 3 + 5 + 7$) and output the sum of the odd numbers.

h. How is the structure of this problem similar to the structure of the problem given in Table 1 (page 88) of this section.

11. Recall the parking lot problem from Chapter 1: A parking lot with twenty spaces is full. There may be both motorcycles and cars in the lot.

a. Let the number of cars in the lot be input and the number of motorcycles in the lot be output. Create a table of five different input/output pairs and draw a function machine for the relationship.

b. Let the number of cars in the lot be input and the total number of wheels in the lot be output. Create a table of five different input/output pairs and draw a function machine for the relationship.

12. Given the following array

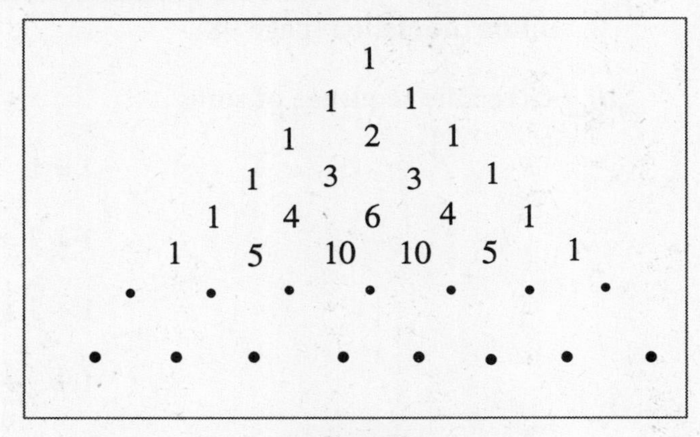

Figure 6

a. Complete another three rows of the array.

b. Sum each row. Describe, in writing, any pattern you discover.

c. List at least three other patterns observed in the array.

d Are there any connections between this array and previous investigations you have done? If so, describe the connections. Identify any patterns.

13. In a school, there are 1000 students and, fortunately, 1000 lockers. The first student comes in and opens every locker. The second student follows and closes every other locker. He is followed by a third student who changes the state (open or closed) of every third locker. The fourth student changes the state of every fourth locker. This process continues until all 1000 students have passed the lockers, each one changing the state of the lockers which match the order in which they pass the lockers. Which lockers are open after all 1000 students have passed the lockers?

14. In your investigation of Rule 1 in the S P C system, how did you use the idea of pattern to decide when to limit your investigation?

15. Each rule in the S P C system is a relation in which the input is an initial path or a valid path. The output is the valid path that results from applying the rule to the input. Which rules in the S P C system are functions? Justify your answer by explaining why certain rules are functions and the remaining rules are not functions. (Hint: See p. 11)

Concept Map

Construct a concept map centered on the word **pattern**.

Reflection

Polya urged students to use the following four steps when attempting to solve a problem:

- Understand the problem.
- Devise a plan.
- Carry out your plan.
- Look back and check your solution.

Classify each of the steps you used in your investigation of Table 1 (page 88) as one of the steps of Polya's problem solving process. What did you do to understand the problem in the S P C investigation? How did you make a plan?

Write a paragraph using your concept map describing the various techniques you can use for problem solving.

Section 3.2 Function: Algebraic Representation

Purpose

- Introduce variable as a quantity that changes as a situation changes.
- Examine behavior of functions using tables.
- Investigate the input–output nature of functions.
- Express functions using equations.

Investigation

Problem Situation: General admission tickets to a concert sell for $19 each. The total receipts (amount of money collected from ticket sales) depend on the number of tickets sold.

1. What would the total receipts be after one ticket is sold? after ten tickets are sold?

2. In Table 1, record the total receipts for the given number of tickets sold. Record the finite difference between successive values for receipts in the space given.

Number of tickets sold	Total receipts
1000	
2000	
3000	
4000	
5000	
6000	

Finite Differences

______

______

______

______

______

Table 1: Concert tickets

3. Given the number of tickets sold, describe in words how to obtain the total receipts.

4. a. What do you notice about the finite differences of the receipts?

   b. As the number of tickets sold increases by 1000, what is the change in the total receipts? Write a complete sentence describing your answer.

5. Record the computation *process* for the total receipts in the right column in Table 2. Use the pattern to write the receipts when n tickets are sold.

Number of tickets sold	Total receipts ($)
1000	
2000	
3000	
4000	
5000	
6000	
n	

Table 2: Concert ticket receipts pattern

6. Let the variable n represent the number of tickets sold and the variable R represent the total receipts.

   a. Draw a function machine representing the relationship.

   b. Write an equation expressing the relationship between n and R.

Discussion

In the above investigations, both a table and an ***equation*** represent the relationship between two changing quantities: the number of tickets sold and the total receipts. In mathematics, variables are used to represent quantities that change as the situation changes. A variable is called an ***independent variable*** if the values of the variable are freely chosen (freely chosen means you make up a number from the list of numbers that are appropriate to the problem). The value of the other variable, the ***dependent variable***, is determined by the choice of value assigned to the independent variable.

Investigation

7. For the concert ticket problem, what quantity represents the independent variable? What quantity represents the dependent variable? Justify your answers.

8. a. What conditions determine the domain of the independent variable?

   b. What are the acceptable values for the independent variable in the concert ticket problem?

Discussion

If the relationship is such that the independent variable determines *exactly one* value of the dependent variable, the relationship between the two variables is expressed using a ***function***.

A function is a process that receives ***input*** (a value of the independent variable) and returns ***output*** (a value of the dependent variable). In the above problem, the input is the number of tickets sold and the output is the total receipts. The set of all legal inputs to the function is called the ***domain of***

the function. The set of all possible outputs of the function is called the ***range of the function***.

The input is a value of the independent variable which is an element of the domain of the function. The output is a value of the dependent variable which is an element of the range of the function.

When discussing the domain and range of a function, we must distinguish between the mathematical function out of context and the ***problem situation***.

If we just consider the function that multiplies an input by nineteen to produce output, then the domain and range is the set of all ***real numbers***. Essentially, any real number can be used as input since all real numbers can be multiplied by nineteen. Any real number can be obtained as output. To see this, we consider how the process is reversed to find the input given the output. In this case, we can reverse the process "multiply by 19" by using the process "divide by 19". Since every real number can be divided by 19, any real number may be output.

Now put the function in context of the problem situation. In this case, the input is the number of concert tickets sold. This must be a whole number less than or equal to the capacity of the concert venue. Thus the domain of the problem situation is the set of whole numbers less than or equal to the maximum number of tickets that may be sold. The output is the total receipts. Since each ticket costs nineteen dollars, the output has to be a whole number multiple of nineteen. Furthermore, it will be less than or equal to nineteen times the capacity of the concert theatre. Thus the range of the problem situation is the set of all whole number multiples of nineteen less than or equal to nineteen times the maximum number of tickets that may be sold.

To present these ideas in another way, Table 3 displays the domain and range of the mathematical relationship versus the domain and range of the problem situation.

Problem	Domain	Range
Mathematical process "multiply by 19".	Set of all real numbers.	Set of all real numbers.
Problem situation: concert tickets sold related to total receipts.	Set of all whole numbers less than or equal to maximum number of tickets.	Set of all whole number multiples of 19 less than or equal to 19 times the maximum number of tickets.

Table 3: Mathematical process versus problem situation

Let's look at ways to represent the function. Table 1 (page 102) displays the function numerically. If n represents the input (the number of concert tickets sold) and R represents the total receipts, Figure 1 displays a function machine for the relationship.

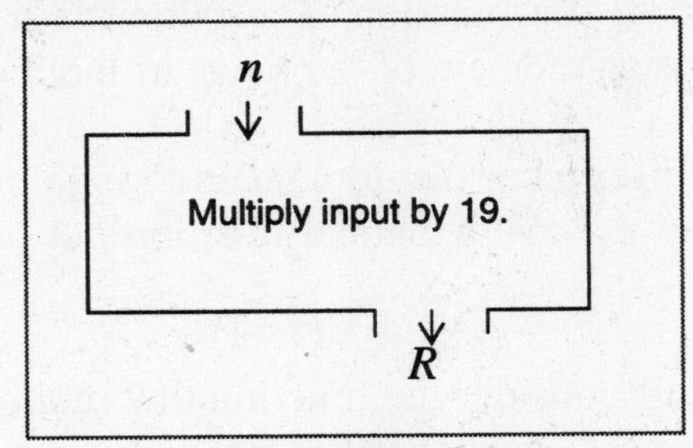

Figure 1

Table 2 (page 103) focuses on the process required to determine the receipts given the number of tickets sold. Once the process is identified, the relationship can be generalized using variables. The generalization is an equation which will be referred to as the algebraic representation of the problem situation.

Table 4 is a completed version of Table 2.

Number of tickets sold	Total receipts ($)
1000	19(1000)
2000	19(2000)
3000	19(3000)
4000	19(4000)
5000	19(5000)
6000	19(6000)
n	$19(n)$

Table 4: Concert ticket receipts pattern

In this case, if n represents the number of tickets sold and R represents the total receipts, the algebraic representation of the problem situation is the equation

$$R = 19n.$$

Notice that the product of 19 and n was represented in Table 4 as $19\,(n)$ and in the equation as $19n$. Both representations are correct. The second, $19n$, is used most commonly when writing the product of a constant and a variable. The constant is written first followed immediately by the variable.

Investigation

9. Use substitution and your calculator to find each of the following for the concert ticket function.

   a. Find the revenue if 500 tickets were sold.

   b. Find the revenue if 1500 tickets were sold.

   c. Find the revenue if 2500 tickets were sold.

   d. Find the revenue if 3759 tickets were sold.

   e. What is n if $R = 114,000$? How did you find n?

Discussion

To answer Investigation 9, we substitute for the appropriate variable in the algebraic representation of the function $R = 19n$.

For example, if 500 tickets were sold, then $n = 500$. So

$$R = 19(500)$$
$$R = 9500.$$

The total receipts when 500 tickets are sold would be $9500.

The above computation may have been done on a calculator in at least three ways.

- Use the calculator to perform the multiplication.
- Use the calculator to store the input and then evaluate the function at that input.
- Use a table to display input/output pairs if your calculator has a table

feature.

The last question states that the total receipts were 114,000. So $R = 114,000$. Substituting into the equation, we get

$$114000 = 19n.$$

To find n, we are finding the input for a given output. This requires us to reverse the process given by the function machine in Figure 1 (page 106). since the process is "multiply by 19", the reverse process is "divide by 19".

To find n, we must divide the output 114000 by 19. So

$$\frac{114000}{19} = n$$

$$6000 = n.$$

If the receipts are 114,000, then 6000 tickets were sold.

Explorations

1. List and define words in this section that appear in ***italics bold*** type using your own words.

2. Evaluate each of the following equations at the given values. Each of these equations represents a relationship between two variables.

   a. If $N = 13b$, find N if $b = 6$.

   b. If $N = 13b$, find b if $N = 936$.

   c. If $C = 4z^3$, find C if $z = 7$.

   d. If $Q = 5m^2$, find Q if $m = 12$.

   e. If $Q = 5m^2$, find m if $Q = 245$.

   f. If $A = 3 + 2b^2$, find A if $b = 13$.

   g. If $A = 3 + 2b^2$, find b if $A = 75$.

   h. If $L = 3r^2 - 2r + 7$, find L if $r = 6$.

3. For each different function in Exploration 2, identify the

   a. independent variable.

   b. dependent variable.

4. For each different function in Exploration 2, describe in words the order of operations required to find the output.

5. Construct a function machine for each different function in Exploration 2.

6. Consider the problem situation: There are six times as many students as professors.

   a. Complete Table 5.

Professors	Students
50	
100	
150	
200	
250	
300	

Table 5: Students versus professors

   b. Complete Table 6 by recording the process required to find the number of students rather than the final computation.

Professors	Students
50	
100	
150	
200	
250	
300	
p	

Table 6: Students versus professors pattern

c. If p represents the number of professors (input) and s represents the number of students (output), use Table 6 to write an equation that algebraically represents the relationship.

d. Using Table 5, find the finite differences between the number of students. What do you notice?

e. If the number of professors increases by 50, how much increase occurs in the number of students? Explain.

f. What is the domain and range of the mathematical process?

g. What is the domain and range of the problem situation?

7. At Harper College, the mathematics department used the equation $S = 72M$, where M is the number of mathematics faculty and S is the number of students, to justify the need for additional mathematics faculty.

   a. State the relationship in words.

   b. If there are 12 full–time mathematics faculty, how many students are enrolled in mathematics courses?

   c. Draw a function machine.

   d. What is the domain and range of the mathematical process?

   e. What is the domain and range of the problem situation?

8. If there are 18 full–time English professors teaching 720 students, write an algebraic representation and draw a function machine for the problem situation.

9. There are only cars parked in a twenty–space parking lot. The lot is not necessarily full. Let the input be the number of cars parked in the lot and the output be the number of wheels in the lot (ignore spare tires).

   a. Write an equation that represents the relationship. Clearly identify the meaning of each of your variables.

   b. For the equation you wrote in part a, which is the independent variable?

   c. Identify the domain and the range of the mathematical process defined in part a.

   d. Identify the domain and range of the problem situation.

10. When we apply a rule in the S P C system to a valid path, we obtain a legal path. This relationship is another example of a relation.

    a. Given the valid path S P P, we obtain the valid path S P P P P using Rule 1. What is the input in this case? What is the output.

    b. Create you own example using a different rule. Indicate the input, the relation or function (rule) used, and the output.

    c. Identify the domain for Rule 1.

    d. Identify the domain for Rule 2.

11. Write your own story problem expressing a relationship between two changing quantities.

    a. Create a table with at least five different rows. Complete the table using the relationship defined in part a.

    b. Write an algebraic representation that defines the relationship you stated in a.

    c. Describe what you were thinking to create the table. Describe what you did to write the equation.

Concept Map

Construct a concept map centered on the word **function**.

Reflection

Use your concept map to write a paragraph describing what a function is. Describe a practical relationship that is an example of a function. Indicate the quantity that represents the input and the quantity that represents the output.

Section 3.3 Function: Notation for Algebraic Representations

Purpose

- Use variables to represent quantities and note how changing one variable causes the other variable to change.
- Discover relationships between quantities in problem situations.
- Introduce function notation.

Investigation

Situation: John has observed that it takes him 25 minutes from the time he wakes up until the time he can leave for school to shower, shave, dress, eat breakfast, etc.

1. List some things in John's morning that are represented numerically.

2. Which of the things are the same each morning (from the information given in the problem)?

3. Which things change?

Discussion

We can use a variable to represent the quantities that change. For example we could let a represent the alarm time when John wakes up and d the time John departs for class. On days he has an 8:00 class, a may have a different value—say 6:30 a.m.—than when he has a ten o'clock class when a might equal 8:00 a.m.

In either case the constant, 25 minutes, can be used to show a relationship between a and d. In words the departure time equals the alarm time plus twenty–five minutes.

The algebraic representation is

$$d = a + 25.$$

Notice that d is dependent on a. In this case a is the ***independent variable*** and d is the ***dependent variable***. This is an example of a ***function***. We will consider a function as a relationship between two quantities that change. In a function, a value of the independent variable (***input***) uniquely determines a value of the dependent variable (***output***).

The set of all inputs to a function is called the ***domain*** of the function. The domain of this problem situation is the set of all possible alarm times.

The set of all outputs of a function is called the ***range*** of the function. The range of this problem situation is the set of all possible departure times.

The relationship $d = a + 25$ can be expressed as a function machine (Figure 1).

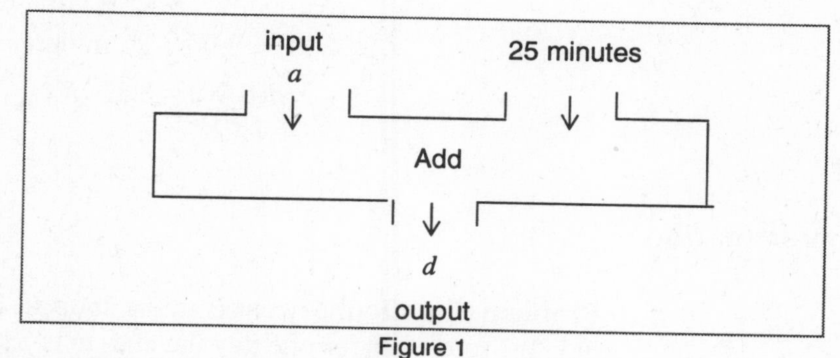

Figure 1

To establish the dependency of d on a, we say that

d is a function of *a*.

The notation $d(a)$ is used as a shortcut for this statement. Be careful to note that this notation does not represent multiplication. It is called ***function notation*** and represents the dependency of d on a. We could then use function notation to rewrite the equation $d = a + 25$ as

$$d(a) = a + 25.$$

Notice that we are using the equal sign to indicate that what is on the left side, $d(a)$, is equivalent to what is on the right side, $a + 25$. Either expression can be substituted for the other when it is convenient to do so.

The general format for function notation is

dependent variable(*independent variable*) = *process*

Alternately, function notation often has the format

function name(*input*) = *output*.

If John wakes up at 6:30 a.m., then we want to find d when a = 6:30. In function notation, this is written d(6:30). In the equation $d(a) = a + 25$ we replace a with 6:30 to get

d(6:30) = 6:30 + 25 minutes (process – do the computation)

d(6:30) = 6:55 (answer or output)

The context determines whether we interpret a + 25 as a process (when we know the value of a and complete the computation) or as an output (no computation, just the representation of the output as a sum).

Similarly, if John wakes up at 8:00 a.m., then we want to find d when a = 8:00. In function notation, this is written d(8:00). In the equation $d(a) = a + 25$ we replace a with 8:00 to get

d(8:00) = 8:00 + 25 minutes

d(8:00) = 8:25

Investigation

Problem Situation: My car is in the shop so I must rent a car for only one day. If I rent a car for one day the charge is $35 plus 17¢ per mile.

4. What are the constants?

5. What are the two variables? Identify the dependent variable. Identify the independent variable.

6. Suppose I drive 53 miles on this day. What will my cost be? Show your work and explain what you are doing.

7. Describe in words how to find the total charge for the car given the number of miles driven.

8. Draw a function machine that demonstrates the relationship.

9. Write an equation to show this relationship mathematically. Use function notation.

10. Numerically determine how many miles I drove if the cost for the day was $48.26. Describe your answer and the process used to find the answer.

Discussion

The relationship that you have written is another example of a ***function***. Notice again that the value of one variable, cost, is determined by the decision of how many miles to drive. If we use the variable m to represent the number of miles driven and the variable C to represent the cost, then m is the input variable and C is the output variable. The function machine appears in Figure 2.

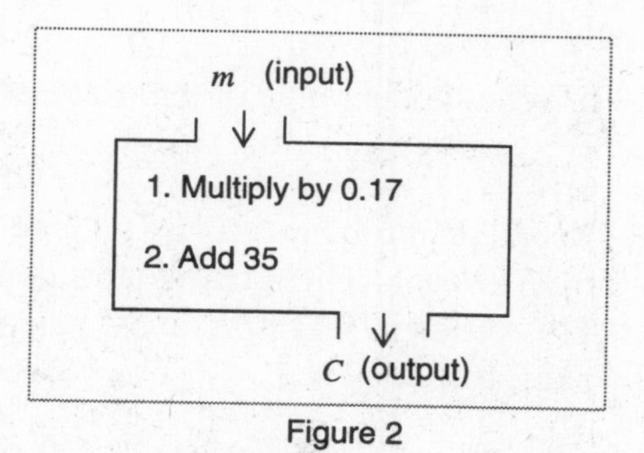

Figure 2

We can express the relationship algebraically as

$$C = 0.17m + 35$$

or we could use function notation and write

$$C(m) = 0.17m + 35.$$

It is important to observe that, in this problem, a variable represents a **number** not an object, so m represents the **number** of miles driven, not miles.

The relations we see between variables will answer such questions as

- how much more is the output than the input
- how many times the input is the output
- what must be done to the input to find the output?

Equations and tables are used to explore these relationships. They can show how, as one variable changes, the other does also–in a predictable way.

Explorations

1. List and define words in this section that appear in ***italics bold*** type using your own words.

2. a. What is the domain of the mathematical process $C(m) = 0.17m + 35$?

   b. What is the domain of the car rental problem situation?

3. a. What is the range of the mathematical process $C(m) = 0.17m + 35$?

   b. What is the range of the car rental problem situation?

4. Recall the problem situation: General admission tickets to a concert sell for \$19 each. The total receipts depend on the number of tickets sold. This problem was first investigated in Section 3.2 beginning on page 102.

   a. Write the algebraic representation of the relationship in function notation.

   b. Select a value for input. Find the corresponding output. Express the results in function notation.

5. Recall the problem situation: There are six times as many students as professors. This problem was first investigated in Section 3.2 on page 109.

   a. Write the algebraic representation of the relationship in function notation.

   b. Select a value for input. Find the corresponding output. Express the results in function notation.

6. Recall the problem situation: Assume there are only cars parked in a twenty–space parking lot. The lot is not necessarily full. Let the input be the number of cars parked in the lot and the output be the number of wheels in the lot (ignore spare tires). This problem was first investigated in Section 3.2 beginning on page 110.

a. Write the algebraic representation of the relationship in function notation.

b. Select a value for input. Find the corresponding output. Express the results in function notation.

7. Write a story problem that expresses the relationship between two changing quantities.

   a. Use the two situations in this section's investigation as a guide.

   b. Identify the input variable and the output variable for the problem in part a.

   c. Draw a function machine for the problem in part a.

   d. Write an algebraic representation (formula) for your problem.

   e. Make up two specific numerical questions that can be answered by using the function you created.

Explorations 8—10 refer to the "mathematical diary" of my day.

Friday

I hate Friday mornings. I'm always tired and there is always so much to do. First I got up late and skipped breakfast so I could leave the house by 7:45. I ran out of gas, but fortunately it was right by a filling station. Unfortunately gas was $1.36 a gallon, but fortunately I had a twenty dollar bill so I could pay for the gas.

By the time I got to class I was twenty minutes late. The students who were still there got an A for the day. Unfortunately that was only half of my class.

I went back to my office and found a light bulb was out. In order to read, I had to move closer to the one remaining bulb. It seemed like there was only one–quarter as much light at two feet away from the bulb as at one foot away. I swear I could only see one–sixteenth as well when I was four feet away. (I must be getting old!)

I had a job interview for the summer. It pays $17 per hour. I wonder if I can work enough hours to equal my current weekly salary.

I was interviewed by six people and it was kind of funny–not only did everyone shake my hand, they all shook each others hand.

On the way home I went through three toll booths. Each time I paid a dollar, but my change was different each time. There must be a lot of different toll charges.

8. Find the quantities that are constant and the quantities that are variables.

9. Indicate variables that have some relation between them and decide which is the input and which is the output.

10. For each pair of variables in Exploration 8, indicate whether increasing the input results in an increase or decrease in the output.

Concept Map

Construct a concept map centered on the word **notation**.

Reflection

A classmate is confused about function notation. Write a paragraph explaining function notation to this person.

Section 3.4 Function: Geometric Interpretations

Purpose

- Introduce graphs as a method of communicating information geometrically.
- Develop the ability to interpret graphs that display relationships between two quantities.
- Develop a geometric representation of functions.

Discussion

Functions can be visualized geometrically using a ***graph***. Graphs are important because they can provide large amounts of information about a function easily. Before we look at functions, we introduce graphs with a game called Mayan Mixup. The game begins with two archeologists, Dr. Art Fact and Dr. Barry Treasure, each having acquired writings describing the location of an ancient Mayan city. Art's information describes the location of the city of the 13th Mayan ruler. Barry's information describes the secret city of the 4th Mayan ruler. Each archeologist is aware of the other's planned search for Mayan treasures. Both Art and Barry have planted spies among the persons working for the other archeologist. Each plans to search for the other's city first, delaying the excavation of his own site to avoid having the other get both treasures first. Each man hopes the other will run out of funds, get tired and leave for home. Based on secret information:

- Each player draws a Reference Map recording the location of **four sites**—the Site of Worship which contains gold and silver artifacts, the Site of Ancient Secrets, and two sites of Mathematical Writings which unlock the secrets of the Mayan culture. (See Figure 1).
- The Site of Worship must intersect 4 ***vertices*** on the reference map, horizontally, vertically, or diagonally. The Site of Ancient Secrets intersects 3 vertices, each site of Mathematical Writings intersects two vertices.
- On each map, vertices are labelled by recording the number of horizontal units from the origin followed by the number of vertical units. The origin (0, 0) is at the lower left corner, the vertex (10, 10) is at the upper right corner. To mark the vertex (3, 7) we move 3 units to the right from (0, 0) followed by 7 units up.
- Art names 7 locations at which he plans to start digging by selecting seven cells in a newly-created Search Plan (See Table 1). He announces the 7 locations by naming the ordered pair corresponding to each of the seven vertices selected. Each of the seven locations is recorded by Art in the Search Plan table with a number 1, identifying the first round of digging locations.

- Barry records each of Art's locations on the appropriate vertices of his map. Barry *says nothing until Art has finished naming locations*, then Barry announces how many sites were located and how many times each site was selected, saying "You located the Site of Worship once" or "You located the Site of Ancient Secrets twice." Barry *does not* name the vertex or vertices of his sites discovered by Art.

- When Art's named location corresponds to a vertex on Barry's map indicating the location of a treasure site, an excavation is recorded. Four hits on Barry's Site of Worship by Art means that Art has recovered all treasure and artifacts from that site. Three hits on the Site of Ancient Secrets and two hits on a each site of Mathematical Writings indicate that those treasures have also been completely excavated.

- If Art locates Barry's treasures, Barry cannot promise his backers a major share of what is recovered and he loses funding as his sites are excavated by Art. The discovery by Art of all treasure and artifacts of a given site also results in fewer attempts at locating sites by Barry. If Barry's Site of Worship is fully excavated by Art, Barry loses 3 location attempts and is reduced to only four attempts to locate Art's treasure sites. Each player loses 2 attempts when his Site of Ancient Secrets is fully excavated and 1 attempt when a Mathematical Writing site is located.

- After Art completes seven location probes, Barry has seven attempts to locate the sites marked by Art, providing none of Barry's sites were completely excavated by Art. Barry records a 1 for each attempt on his Search Plan. Art records the attempts taken by Barry on his map. Play continues until one player has located all of the opponent's sites.

Investigation

Begin the game by creating a reference map of your four sites. Record the vertices where each site is located.

Site of Worship:________________________________

Site of Ancient Secrets: ____________________________________

Math Site 1: _________________ Math Site 2: _________________

1. On your Search Plan, select and record the vertices where you think the opponent's sites are located. Note the vertices you locate.
2. As your opponent names the vertices where he/she thinks your sites are located, record those vertices on your reference map.
3. Complete Round 1 by trying to locate the sites of your opponent.

4. Complete the game. Keep track of all attempts and successful probes in each round, using the Search Plan and Reference Map below.

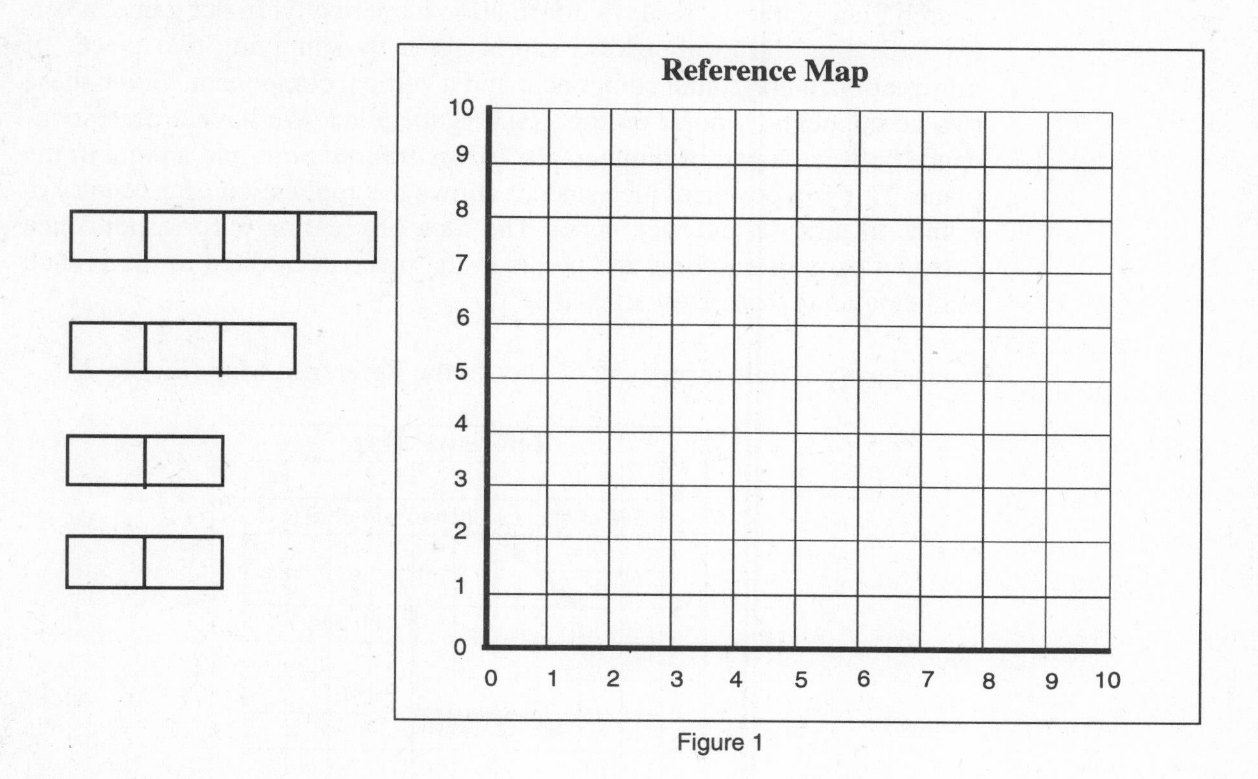

Figure 1

10											
9											
8											
7											
6											
5											
4											
3											
2											
1											
0											
	0	1	2	3	4	5	6	7	8	9	10

Table 1: Search Plan

Discussion

The Reference Map (Figure 1) displays information visually while the Search Plan (Table 1) displays information numerically. In our game, a vertex indicating the location of a site is given by supplying two pieces of information: a horizontal component and a vertical component. Given these two components, a point on the graph is specified. We have a correspondence between a pair of numbers, called an ***ordered pair***, and a point in the plane. This is a powerful idea since it allows the application of geometry (a graph) to algebra and vice versa. The idea of creating a correspondence between pairs of numbers and points in the plane is credited to the French mathematician Renè Descartes in 1637.

Consider a possible placement of sites in the Reference Map (Figure 2).

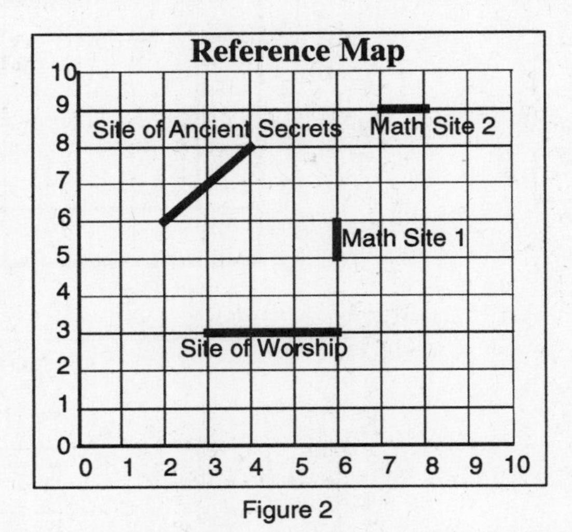

Figure 2

We label the location of each site using pairs of numbers for the vertices and enclose each ordered pair in parentheses, separating the numbers by commas, listing the horizontal component first. The Reference Map in Figure 2 has the following locations:

- Site of Worship: (3, 3), (4, 3), (5, 3), (6, 3).
- Site of Ancient Secrets: (2, 6), (3, 7), (4, 8).
- Mathematical Writings Site1: (6, 5), (6, 6).
- Mathematical Writings Site 2: (7, 9), (8, 9).

When considering functions, we also work with pairs of numbers. For any function, we have an input number that uniquely determines the output number. We could look at these as ordered pairs of the form (input, output) (See Figure 3).

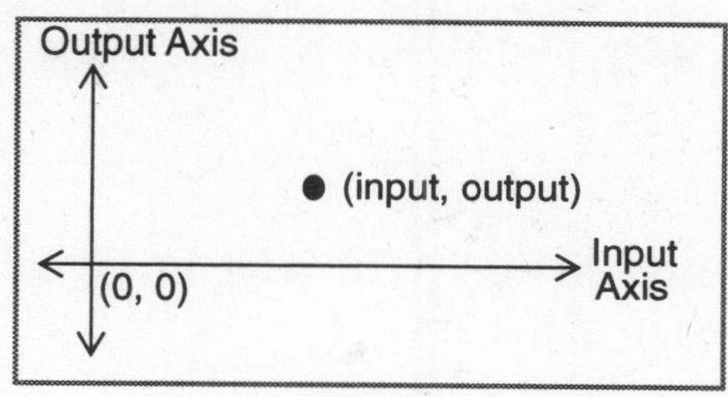

Figure 3

Each such pair could then be graphed just as we did with the sites. We have created a geometric view of the function.

To create such a graph we first generate ordered pairs by recording input–output pairs. We then draw two number lines, called ***axes***, representing the horizontal and vertical components. The number lines meet at the point $(0, 0)$ called the ***origin***. We next mark off each number line with ***tick marks*** representing the units on each axis. It is not necessary for both axes to use the same units between tick marks. In the Search Plan the distance between tick marks was one unit on both axes. Finally we ***plot*** ordered pairs by locating the pairs as points on the graph. This creates the graph of the function.

Investigation

Let's create a graph based on a practical example. In a previous section, we considered the following situation: General admission tickets to a concert sell for $19 each. From this, we created the Table 2.

Number of tickets sold	Total receipts
1000	19000
2000	38000
3000	57000
4000	76000
5000	95000
6000	114000

Table 2: Concert ticket receipts.

5. Refer to Table 2:

   a. Name the input quantity.

   b. Name the output quantity.

6. Express the information in Table 2 as ordered pairs.

Investigations 7 - 10 refer to Figure 4 where we graph the information given in Table 2.

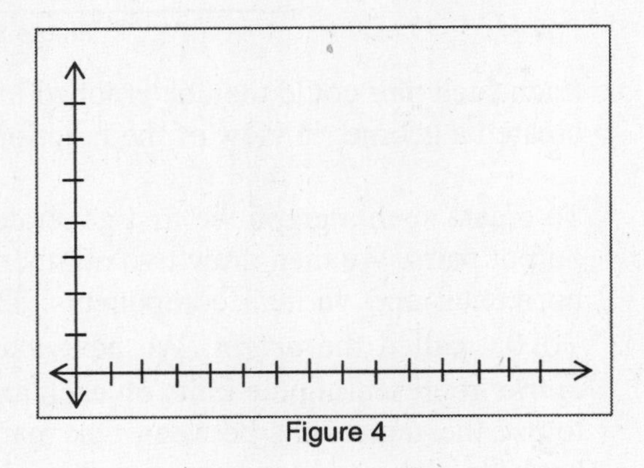

Figure 4

7. Use words to label the input ***axis*** and output ***axis***.

8. Record reasonable values below the tick marks on the horizontal axis in order to represent the input.

9. Record reasonable values to the left of the tick marks on the vertical axis to represent output.

10. Mark and label a point for each ordered pair specified by Table 2.

Discussion

We have now investigated the concert ticket relationship in three ways.

- Numerically using Table 2.
- Algebraically using the equation $R(n) = 19n$.
- Geometrically using the graph constructed in Figure 4. In Figure 5 you

see a graph of the ordered pairs given in Table 2.

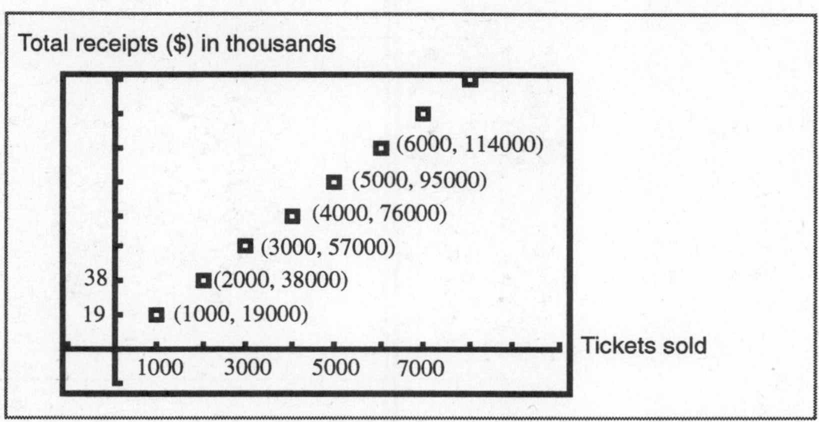

Figure 5

As we study relationships involving two varying quantities in this course, we will consider all three representations. Each has its strong and weak points in helping us understand the relationship.

Explorations

1. List and define words in this section that appear in ***italics bold*** type using your own words.

2. Given the Reference Map (Figure 6), identify the locations of each site by recording the ordered pairs of vertices where each site has been located.

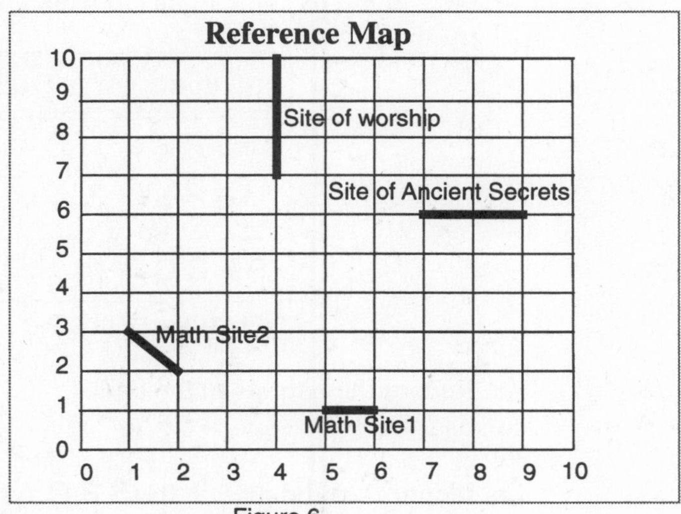

Figure 6

3. On the grid given in Figure 7 graph the ordered pairs (30, 2), (50, 7), (10,9), (60, 0), (35, 6), and (0,4).

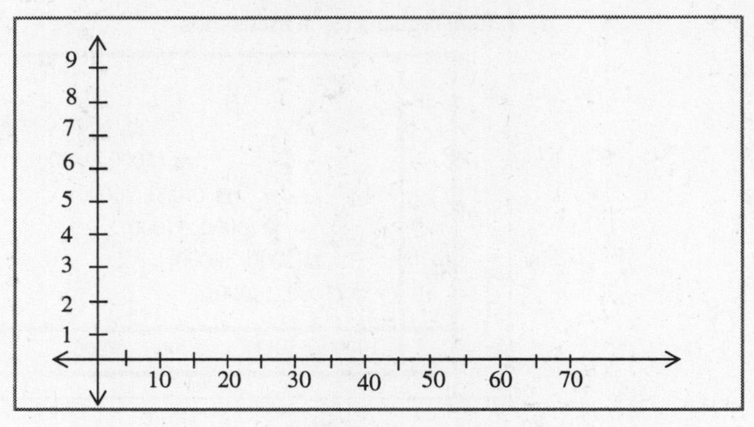

Figure 7

In Explorations 4 and 5, look at previously–studied relationships from a geometric (graphical) viewpoint. For each of these explorations:

a. Create a table of at least six input/output pairs.

b. Write the algebraic representation.

c. Graph the relationship.

4. There are six students for every professor at the university.

5. The input is a whole number and the output is the cube of the whole number.

6. Table 3 contains some inputs to Rule 2 in the S P C system.

Input	Output
S P P	
S C P	
S P C C P	

Table 3: Rule 2 of S P C

a. Record the outputs in the table.

b. Identify a valid path in the S P C system that is not in the domain of the function defined by Rule 2. Explain.

7. Three graphs are drawn on the axes shown in Figure 8. Match each graph with the situation best depicted by it. Justify your answer.

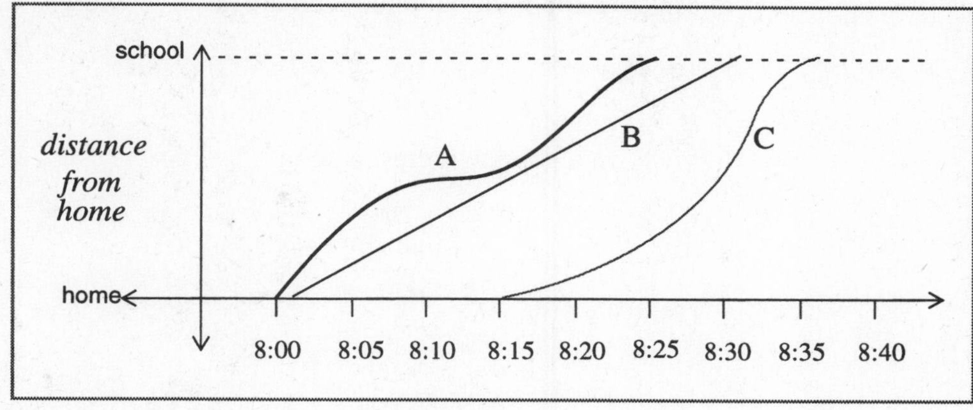

Figure 8

a. Sandy walked to school.

b. Joe rode his bike to school.

c. Jane left late and had to run part of the way to school.

d. What could be the cause of graph A's being level for a few minutes between 8:05 and 8:10?

8. Identify all prime numbers between 90 and 100. Express your answer as a set.

9. Is the set of prime numbers less than 100 closed under the operation of addition? Justify your answer.

10. Identify three different ways to use a variable. Give an example of each.

11. Use F for feet and Y for yards. Let yards be the input and feet be the output.

    a. Make an input/output table with at least five entries.

    b. Write an algebraic representation.

    c. Draw a graph.

12. Write an example of an equation.

Concept Map

Construct a concept map centered on the word **graph**.

Reflection

Discuss the positive and negative aspects of representing a relationship between two varying quantities using

a. a table.

b. an equation.

c. a graph.

How do you see these representations fitting together?

Section 3.5 Triangular Numbers

Purpose

- Explore triangular numbers numerically and algebraically.
- Discover relationships between quantities in problem situations.

Investigation

Problem Situation: Suppose you are offered a job in which you will be paid $1 the first day, $2 the second day, $3 the third day, etc. Naturally, you want to keep track of how much money you will have at the end of two days, nine days, or a year, and to compare this way of being paid with other, more traditional ways of being paid. Visually representing the problem can be an efficient and effective way to understand the problem. In the illustration below, each dot represents a dollar, each row represents one day's wages, and each grouping of dots represents the money earned to date.

1. The pattern of four triangles in Figure 1 visually displays the total amount of money earned after one, two, three, and four days.

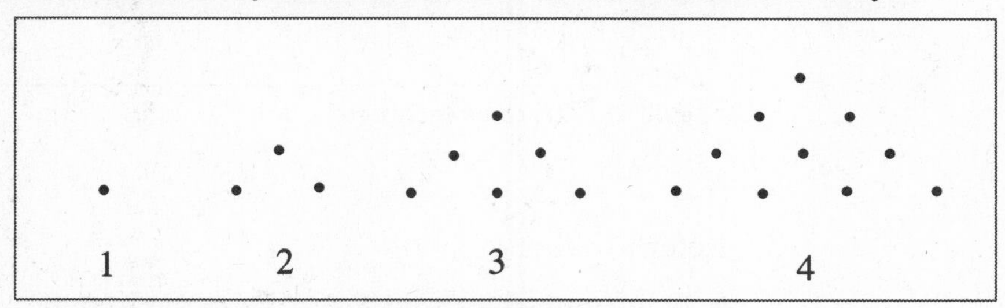

Figure 1

a. Draw the next two triangles in the ***sequence***.

c. How many dots were added to the fourth triangle to generate the fifth triangle? How many dots were added to the fifth triangle to generate the sixth triangle? How does this relate to the amount paid each day?

Draw a picture to understand the problem better.

Problem Solving Toolkit

2. Based on Figure 1 and your answers to Investigation 1, complete Table 1.

Day	Day's wages	Total wages to date expressed as an indicated sum	Total wages to date
1	1	1	1
2	2	1 + 2	3
3	3	1 + 2 + 3	
4			
5			
6			
7			
8			
9			
10			

Table 1: Pattern of earnings

3. Refer to Table 1.

   a. How do the day's wages relate to the triangles in Figure 1?

   b. How does the third column relate to the triangles in Figure 1?

   c. How does the fourth column relate to the triangles in Figure 1?

   d. Describe how you calculated the answers in the fourth column.

4. Refer to Table 1.

   a. How much money is earned on the twentieth day?

   b. Express the total amount of money earned after twenty days as an indicated sum similar to column 3.

   c. If you knew the total wages after nineteen days, how would you calculate the total wages after twenty days?

   d. How much money is earned on the one–hundredth day?

   e. Express the total amount of money earned after one hundred days as an indicated sum similar to column 3.

   f. How much money is earned on the nth day where n is any whole number?

   g. Express the total amount of money earned after n days, where n is any whole number, as an indicated sum similar to column 3.

Discussion

We have investigated several functions in the study of the wages problem situation. The simplest of these is the relationship between the day number and the wages earned that day. If you compare the amounts in columns 1 and 2 of Table 1, you find that they are the same. The day acts as input and the day's wage acts as output. Thus, the output is identical to the input. This function is called the ***identity function***. We represent the identity function using function notation as

$$I(n) = n.$$

Figure 2 displays a function machine for the identity function.

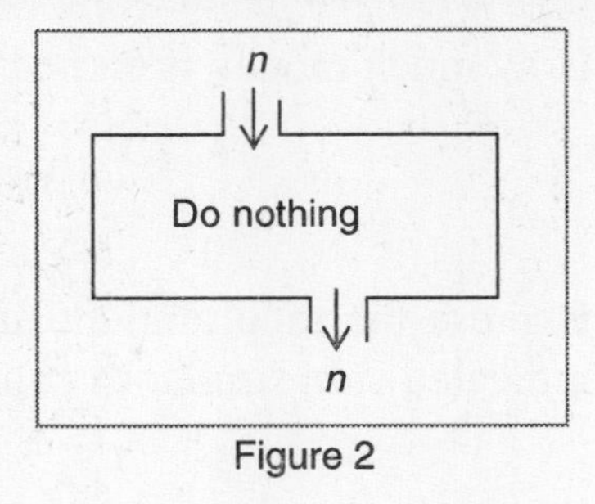

Figure 2

Notice that this function also expresses the relationship between the triangle number in Figure 1 and the number of dots in the bottom row of the triangle.

A second function is the relationship between the amount earned in a given day and the total amount earned expressed as an indicated sum. In this case the input is a whole number and the output is the indicated sum of all whole numbers less than or equal to the input. For example, if we earn eight dollars on a given day then we have earned $1+2+3+4+5+6+7+8$ thus far. Let *IS* (indicated sum) be the name of this function. The above statement expressed in function notation is

$$IS(8) = 1+2+3+4+5+6+7+8.$$

If we assume n represents any whole number, we may generalize the function as

$$IS(n) = 1+2+3+4+\ldots+n.$$

Figure 3 displays a function machine for this the *IS* function.

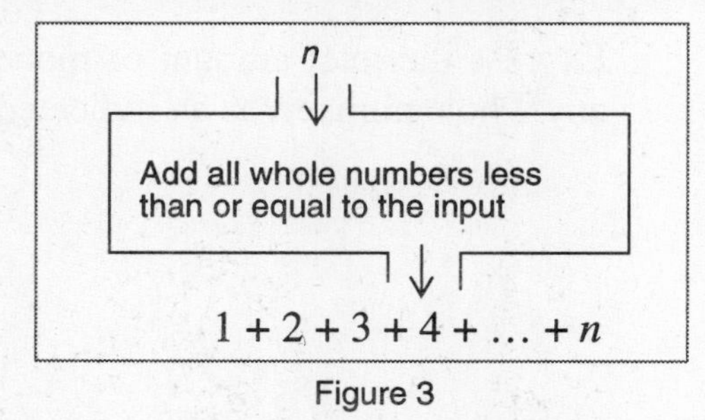

Figure 3

Notice this function also expresses the relationship between the number of dots in the bottom row of the triangle in Figure 1 and the indicated sum of the number of dots in all rows of the triangle.

The next function we consider is the one with input the indicated sum of consecutive whole numbers and output the computed sum. For example, given the indicated sum $1+2+3+4+5+6+7+8$, the computed sum is 36. Let *CS* (computed sum) be the name of this function.

The previous statement can be expressed as

$$CS(1+2+3+4+5+6+7+8) = 36.$$

If we assume n represents any whole number, we may generalize the function as

$$CS(1+2+3+4+\ldots+n) = ?.$$

The question mark on the right hand side states that we don't know how to calculate this yet. Read on and maybe we can figure out a formula for this function. We'll need such a formula if we want to know how much money we have made after one hundred days or one thousand days.

Figure 4 displays a function machine for this function.

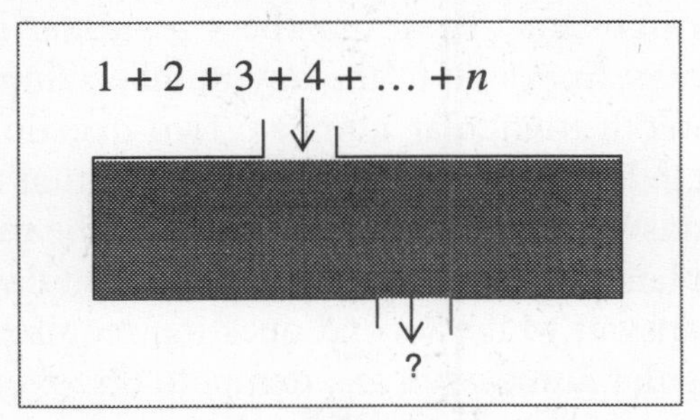

Figure 4

The inside of the function machine in Figure 4 is shaded hiding the process from us. We'd like to figure out what this process is.

It's time to put all this together. Consider the function with input the day number and output the total amount of money earned by the end of that day. Another way to describe this function is to think of input as the triangle number in Figure 1 and output the total number of dots in the triangle. If we use on the first and last columns of Table 1, we have a numerical representation of this function (Table 2).

Day	Total wages to date
1	1
2	3
3	6
4	10
5	15

Table 2: Day versus total earned

Day	Total wages to date
6	21
7	28
8	36
9	45
10	55

Table 2: Day versus total earned

The sequence of numbers in the output column of Table 2 forms a common set of numbers in mathematics. The set is named the ***triangular numbers***. Do you see why? The output column of Table 2 contains the first ten triangular numbers. These numbers represent ***cumulative sums*** of the whole numbers from one to the day number. Suppose we want to know the one hundredth triangular number. Two questions follow: Why would anyone want to know the one hundredth triangular number? How could we find it? The answer to the first question would be relevant depending upon whether you plan to quit work after one hundred days or perhaps buy a new stereo. The answer to the second question requires us to analyze the ***sequence*** of triangular numbers in an attempt to discover a method for easily generating any triangular number. If we discover such a method, then we can state the process for the function machine in Figure 4.

Ultimately we want a function that begins with the function in Figure 1 and concludes with the function in Figure 4. Figure 5 displays such a function.

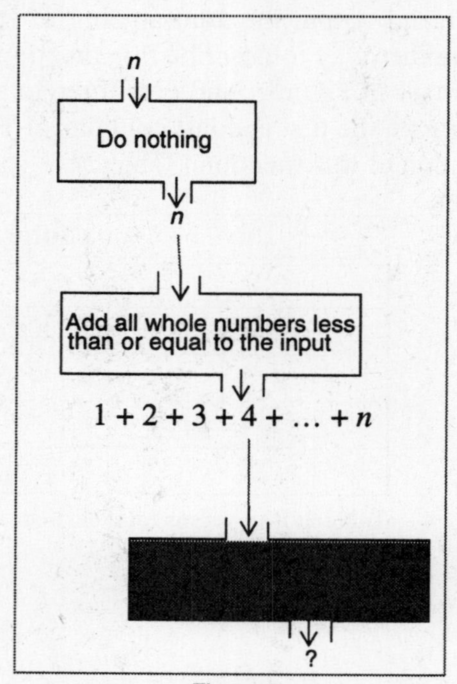

Figure 5

In order to talk about a specific triangular number we refer to it's location in the sequence. The position (location) number is referred to as an ***index***.

Index	1	2	3	4	5	6	7	8	9	10
Triangular Number	1	3	6	10	15	21	28	36	45	55

Table 3: Triangular number position

Notice that Table 3 is another version of Table 2. Matching a triangular number with its index provides us with the means to represent any triangular number, using variables to represent the index and the corresponding triangular number.

Let T represent any triangular number and n its position (index) in the sequence. The index for the first triangular number is one and can be written as $n = 1$. The corresponding triangular number, represented by the variable T, is also 1, written $T = 1$.

The fourth triangular number has an index of 4 ($n = 4$) and the fourth triangular number is 10 ($T = 10$).

We generated the tenth triangular number by adding up the first ten counting (natural) numbers. The hundredth triangular number would be the sum of the first one hundred counting numbers. It would be a tedious process to find any triangular number if we had to add up the first n counting numbers every time we wanted to find another triangular number. To find the ninth triangular number we need to add the first nine counting numbers:

$$1 + 2 + 3 + 4 + 5 + 6 + 7 + 8 + 9$$

To find the hundredth triangular number, we would need to find the sum

$$1 + 2 + 3 + 4 + \ldots + 97 + 98 + 99 + 100.$$

Finding the sum of the first nine counting numbers is tedious. Finding the hundredth triangular number by adding up the first one hundred counting numbers is a task we would not want to do. Neither did a young boy named ***Carl Friedrich (Freddy) Gauss***. He had a clever idea that saved time and energy. (We wish we had thought of the idea first).

Investigation

5. Consider the sequence of the first ten counting numbers in Figure 6.

a. Add the pairs of numbers connected by the dashed lines. Record your answers on the dashed lines.

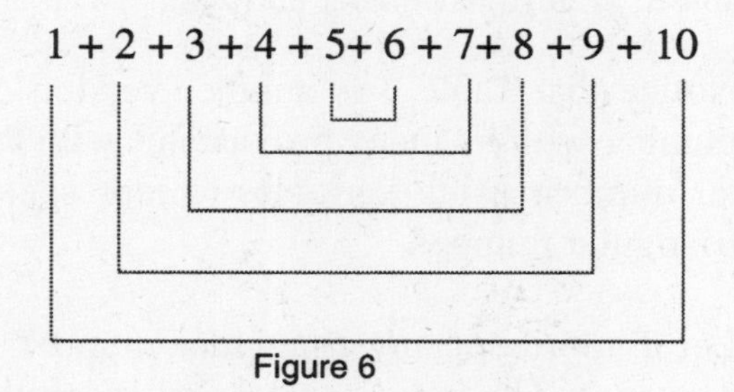

Figure 6

b. What is the sum of each pair of numbers from Figure 6? How could you use the last number to obtain this answer?

c. How many pairs of numbers are there? How could you determine this number just using the last number (10) in the sum?

d. Find the sum of the first 10 counting numbers using your answers to parts b. and c. Which triangular number have you computed?

6. Find the sum of the first twenty counting numbers using the same technique that was used in Investigation 5. Which triangular number have you computed?

7. Find the sum of the first one hundred counting numbers using the same technique as in Investigation 5. Which triangular number have you computed?

8. Write an algebraic expression for the sum $1 + 2 + 3 + 4 + \ldots + n$. As an aid refer to the approach used in Investigations 5–7.

Discussion

To calculate $1 + 2 + 3 + 4 + 5 + 6 + 7 + 8 + 9 + 10$, we add the first and last numbers to get 11. We multiply 11 by one–half of the last number. We use one–half the last number since we are taking the numbers in pairs. Thus

$$1 + 2 + 3 + 4 + 5 + 6 + 7 + 8 + 9 + 10 = \frac{1}{2}(10)\ (11) = 55.$$

The tenth triangular number is 55. Also, the sum of the first ten counting numbers is 55.

To calculate $1 + 2 + 3 + 4 + \ldots + 20$, we add the first and last numbers to get 21. We multiply 21 by one–half of the last number.

$$1 + 2 + 3 + 4 + \ldots + 20 = \frac{1}{2}(20)\ (21) = 210.$$

The twentieth triangular number is 210. Also, the sum of the first twenty counting numbers is 210.

To calculate $1 + 2 + 3 + 4 + \ldots + 100$, we add the first and last numbers to get 101. We multiply 101 by one–half of the last number.

$$1 + 2 + 3 + 4 + \ldots + 100 = \frac{1}{2}(100)\ (101) = 5050.$$

The one hundredth triangular number is 5050. Also, the sum of the first one hundred counting numbers is 5050.

Now let's generalize. To calculate $1 + 2 + 3 + 4 + \ldots + n$, we add the first and last numbers to get $n + 1$. We multiply $n + 1$ by one–half of the last number.

$$1 + 2 + 3 + 4 + \ldots + n = \frac{1}{2}(n)\ (n + 1).$$

By completing Investigation 8 you have developed a process to generate any triangular number (and which also finds the sum of any n counting numbers). We express this process as a formula. Formulas are used to express relationships between changing quantities. In this case, as the index n changes, the triangular number T changes.

- n is the variable representing the index (position) in the triangular numbers sequence and also represents how many counting numbers we wish to add. We can represent the problem geometrically by drawing a triangle with n dots on a side.

- T is the variable name of any triangular number and the sum of the first n counting numbers. Geometrically, the sum of the first n counting numbers or the triangular number is represented by the total number of dots in the triangular array with n dots on a side.

Adding pairs of numbers in a sequence is a way of finding the sum of the first n counting numbers in order to develop a formula for finding the nth triangular number (another name for the sum of the first n counting numbers). In the pairing process, recognizing that we have half the number of sums as there are numbers in the sequence, we divide n by 2, then multiply the quotient by $n + 1$. The formula for finding the sum of the first n counting numbers, or the nth triangular number can also be written

$$T = \frac{1}{2}n(n+1).$$

Since T is a function of n, we can write the formula in function notation as

$$T(n) = \frac{n(n+1)}{2}.$$

A function machine for this function appears in Figure 7.

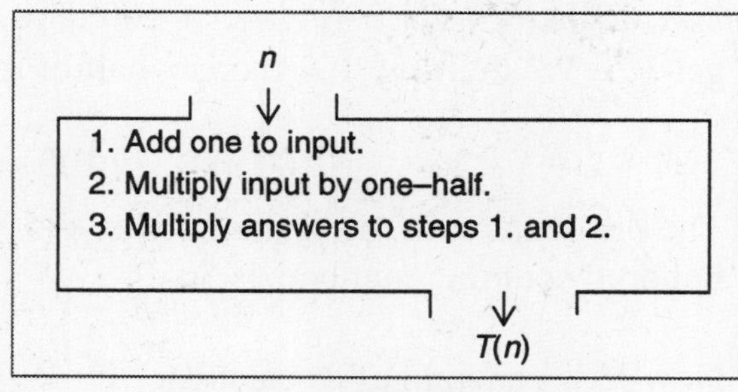

Figure 7

The notation $T(n)$ represents the output, a triangular number, when n is input. The process of replacing n with a whole number is called ***substitution***.

If $n = 4$, every n in the equation is replaced by 4. The equation becomes

$$T(4) = \frac{1}{2}(4)(4+1) = 2(5) = 10.$$

This says that the fourth triangular number is ten.

Finally, we can complete the function machine originally displayed in Figure 5. Recall that the last function machine originally was shaded because we did not know how to sum the first n natural numbers. Now we can use the process from Figure 7 to complete the last process.

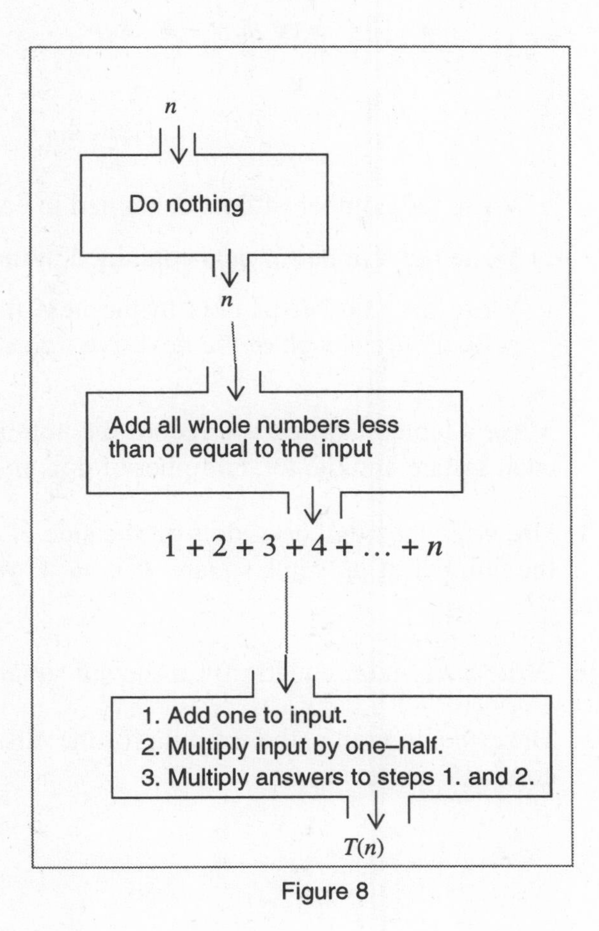

Figure 8

Explorations

1. List and define words in this section that appear in ***italics bold*** type using your own words.
2. Use the formula $T = \frac{1}{2}n(n+1)$ to find T if $n = 7$. Does your answer agree with the value in the table?
3. Does the formula work for odd numbers? Explain.

4. Find the number of dots in a triangle with 113 dots on a side. Write down the algebraic notation with your answer.

5. If $n = 2173$, find T. Show your work.

6. Given the pattern of four squares in Figure 9.

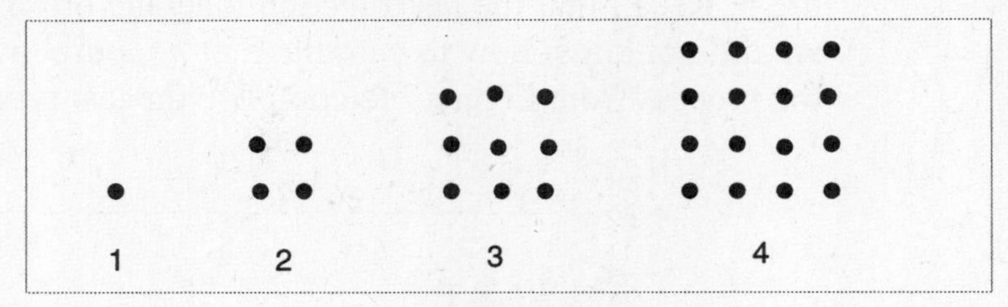

Figure 9

a. Write the number of dots contained in each square.

b. Write the number of dots contained in the next square.

c. Write the number of dots in the next five figures. Did you need to actually draw each of the next five figures?

7. Make a table in which you record the number of dots on the side of each square and the total number of dots in each of the ten squares.

8. Use n for the number of dots on the side of a square and S to represent the number of dots in a square. If $n = 4$, what is S? If $n = 8$, what is S?

9. Write a formula (equation) for the nth square number.

10. Describe, in words, the formula for the nth triangular number. Use complete sentences.

Concept Map

Construct a concept map centered on the phrase **triangular number**.

Reflection

Use your concept map to write a paragraph describing the triangular numbers.

Section 3.6 Power and Factorial Functions

Purpose

- Represent the class of power functions numerically and algebraically.
- Develop understanding of the factorial function numerically and algebraically.

Investigation

1. Complete each of the following tables. Compute the finite differences in the outputs. Then compute the second finite differences by finding the difference in the first finite differences. Continue to take finite differences until they become constant.

Whole number	Cube of whole number
1	
2	8
3	
4	
5	
6	
7	
8	

Table 1: Cubes

Finite differences

___ ___ ___ ___ ___ ___ ___

___ ___ ___ ___ ___ ___

___ ___ ___ ___ ___

___ ___ ___ ___

Whole number	Fourth power of whole number
1	
2	16
3	
4	
5	
6	
7	
8	
?	10000

Table 2: Fourth powers

Finite differences

Whole number	Fifth power of whole number
1	
2	32
3	
4	
5	
6	
7	
8	
?	100000

Table 3: Fifth powers

Finite differences

2. Earlier in the text, you wrote $S(n) = n^2$ for the squares of the whole numbers.

   a. For Tables 1–3, state the variables you will use for input and for output.

      Table 1:

      Table 2:

      Table 3:

   b. For Tables 1–3, describe in words how to compute the output given a value for input.

      Table 1:

      Table 2:

      Table 3:

   c. For Table 1–3, draw a function machine.

      Table 1:

      Table 2:

      Table 3:

   d. For Tables 1–3, write an algebraic representation for the function.

      Table 1:

      Table 2:

      Table 3:

3. For Tables 1–3, record below the number of finite differences required until the finite differences become constant.

   Table 1:

   Table 2:

   Table 3:

4. How could you determine the function from the number of finite difference required to get a constant? Write a complete sentence stating your answer.

Discussion

In the above investigations, we have looked at three different members of a class of functions called ***power functions***. In addition, you looked at another member of the class, the squaring function, earlier in the text. These functions represent an important class of functions in which the input variable is raised to a specific power. This general class can be represented by the definition

$$nPower(x) = x^n$$

Figure 1 demonstrates a function machine for this important class of functions.

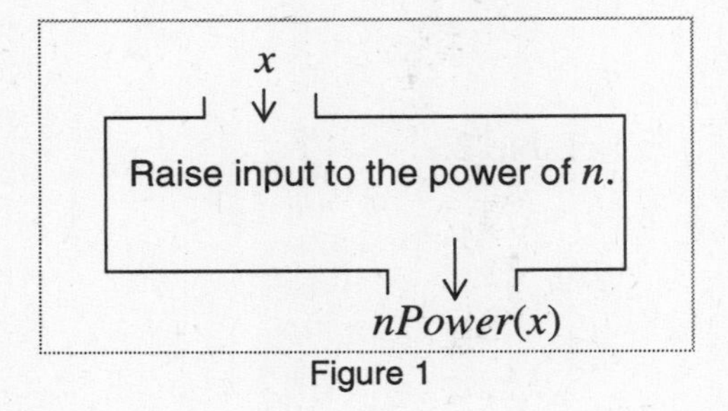

Figure 1

The letter n will be some whole number greater than 0. Once a value for n is chosen, a specific power function is indicated. For example, if $n = 3$, then the power function is the cubing function. In all cases, x acts as the input.

We should notice that the finite differences for each power function eventually become constant. The number of differences needed before they become constant is closely related to the power on the function. This fact will be very important when we attempt to create functions for given data.

Investigation

5. There is an English book and a Chemistry book standing side by side on a shelf.

   a. List the possible arrangements of the two books on the shelf.

   b. How many ways can you arrange them?

6. Your roommate returns your French book.

   a. List the possible arrangements of the three books.

   b. How many ways can you arrange them?

7. You finally finish your math assignment and put your math book on the shelf with the three other books.

   a. List the possible arrangements.

   b. How many ways can you arrange them?

   c. Describe any patterns you have followed to make sure you didn't miss any.

8. Predict the number of arrangements of 5 books. Do you want to make the list?

9. List one idea for finding the number of arrangements. Look back at your previous answers.

10. In your group, decide which idea you will use first to solve this problem.

11. Use your answers to Investigations 5 - 10 along with any pattern you've observed to complete Table 4.

Number of books	Number of arrangements
1	
2	
3	
4	
5	
6	
7	

Table 4: Arrangements of books

12. Describe how you could use Table 4 to find the number of arrangements with 8 books.

Discussion

We have just investigated a very important function, called the ***factorial function***. In this case the input to this function is a whole number of objects. The output is the number of ways we can arrange these objects given by the input. We will represent this function by $Fact(n)$.

Example: If $n = 8$, $Fact(8) = (8)(7)(6)(5)(4)(3)(2)(1)$.

So Fact(8) = 40,320.

This function is defined as the product of the whole numbers between n and 1 inclusive. $Fact(n)$ is represented as $n!$. The function machine appears in Figure 2.

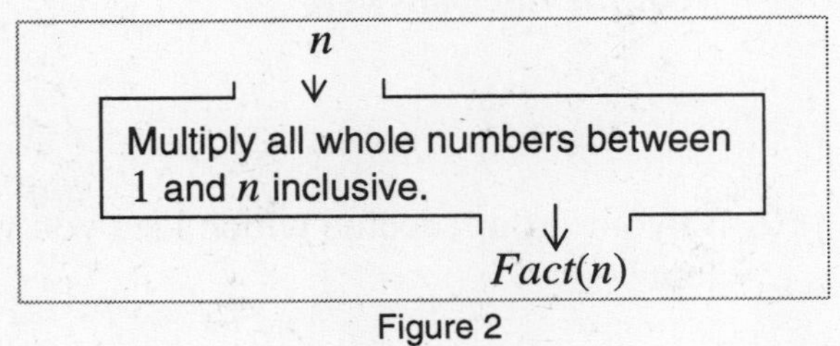

Figure 2

Example: $Fact(8)$ is written as $8!$ So $8! = 40,320$.

Many calculators have a factorial key. Does yours? Ask your instructor if you are unsure. The factorial function is very important when we want to count arrangements. This function is used when we want to calculate the chances of winning the lottery, for example.

Explorations

1. List and define words in this section that appear in ***italics bold*** type using your own words.

2. Ten people are in a room. At least one person is a student and, in any group of three people in the room, at least one is not a student. How many students are in the room? Describe how you found your answer.

3. Sometimes data is collected and the problem is to write a mathematical representation that shows the relationship. A very nervous skydiver on her first jump performs the following experiment to determine the meaning of the phrase "fall like a rock". She drops a very stale corned beef sandwich from a high flying plane. Table 5 shows the relationship between the distance (in feet) the sandwich has fallen and the time (in seconds) since it was dropped.

Time (seconds)	Distance fallen (feet)
1	16
2	64
3	144
4	256
5	400
6	576
7	784
8	1024
9	1296
10	1600

Table 5: The travels of a stale corned beef sandwich

a. One way to look for a pattern is to use finite differences. Find the finite differences between the distances fallen. When do they become constant?

b. Based on your answer to part a, will the function representing the relationship between distance fallen and time be based on squares, cubes, fourth powers, or fifth powers? Why?

c. If $D(t)$ represents the distance fallen during the first t seconds, write an algebraic representation that might define this relationship. You can check your conjecture by substituting a time value from the table, doing the computation, and checking to see if the answer comes out to be the corresponding distance.

4. There are nine players and nine positions on a Little League baseball team.

a. How many different arrangements are there of the players in the nine positions?

b. If a different arrangement of players in positions is tried each inning, how many innings are necessary to try all the possible arrangements?

c. If each game lasts 6 innings, how many games are required to try all possible arrangements? (Assume a new arrangement is used each inning.)

d. If one game is played each day, how many years are required to try all possible positions. (Assume a new arrangement is used each inning.)

Concept Map

Construct a concept map centered on the word **factorial**.

Reflection

Describe two real–life relationships that are examples of a function. For each indicate the quantity that represents the input and the quantity that represents the output.

Section 3.7 Making Connections: What is a Function?

Purpose

- Reflect upon ideas explored in Chapter 3.
- Explore the connections among representations of functions.

Investigation

In this section you will work outside the system to reflect upon the mathematics in Chapter 3: what you've done and how you've done it.

1. State the five most important ideas in this chapter. Why did you select each?

2. Identify all the mathematical concepts, processes and skills you used to investigate the problems in Chapter 3?

3. What has been common to all of the investigations which you have completed?

4. Select a key idea from this chapter. Write a paragraph explaining it to a confused best friend.

5. You have investigated many problems in this chapter.

   a. List your three favorite problems and tell why you selected each of them.

   b. Which problem did you think was the most difficult and why?

6. How does the use of function notation make it easier to understand functions? What is confusing about function notation?

Discussion

There are a number of really important ideas which you might have listed including function, function machines, function notation, triangular numbers, and algebraic, geometric, and numeric representations of situations or expressions.

Concept Map

Construct a concept map centered around the phrase **function representation.**

Reflection

What have you learned by looking at the triangular numbers visually, numerically, graphically and algebraically?

Illustration

Draw a picture of **a mathematics student** learning about functions.

A MATHEMATICS
STUDENT

Chapter 4

Integers: Expanding on a Mathematical System

Section 4.1 Integers and the Algebraic Extension

Purpose

- Introduce the opposites of the whole numbers.
- Investigate signed number operations.

Discussion

In any type of golf game (regular, miniature, frisbee, etc.), the score on a hole is often measured by its signed distance from par—the "expected" score on a hole. For example, a "birdie" represents one shot below par (–1) and a "bogey" represents one shot over par (+1). Par is represented by 0. The person with the lowest score has performed the best. Let's investigate positive and negative numbers by playing "calculator golf".

You will play a nine–hole game. The pars for each hole appear in Table 1.

Hole	1	2	3	4	5	6	7	8	9
Par	4	5	4	3	5	4	4	3	4

Table 1: Calculator golf course

You will use the random number generator on your calculator to "play" each hole. To do so, enter the expression

$$Int(Rand \text{ x } 8) + 1$$

in the Home Screen. Your instructor can help if you are unsure how to do this. The meaning of the expression is explained in the next discussion.

Investigation

1. After entering *Int*(*Rand* x 8) + 1 in the Home Screen, press **Enter**. The result is your score for the first hole. Write this score down. Also record the signed score as measured from par.

2. For the first hole, who scored the best in your group? Record the scores as measured from par of each member in your group from best to worst.

3. Record your answers to Investigation 1 in Table 2. Press **Enter** to continue playing the game. The score for the second hole appears. Press **Enter** seven more times to determine your score on the next seven holes. Record your results in Table 2.

Hole	1	2	3	4	5	6	7	8	9
Par	4	5	4	3	5	4	4	3	4
Raw score									
Signed score as measured from par									

Table 2: A complete game of calculator golf

4. Determine the best overall score in your group. Record this score as measured from par on the board or overhead.

5. After all the top scores are recorded on the board, order them beginning with the best score. Record the ordered list.

You previously completed nine holes. A complete round consist of eighteen holes. In Investigation 6, you will consider the score as measured from par that you would need on the second nine in order to play par golf for a complete round.

6. In the first column of Table 3 record the scores as measured from par of those in your group and one other group. In the second column, record the scores that are opposites. These are the scores for the next nine holes so that the person's overall score would be equal to par or 0. In the third column, record the sums of column one and two.

Scores as measured from par	Opposites of column one	Sum of columns one and two

Table 3: Opposites

7. As a result of Investigation 6, write a complete sentence describing the sum of a and $-a$.

8. Review the section that introduces whole numbers. The whole numbers were not closed under two operations. What were the two operations? Give an example for each showing why.

9. Given a number system that includes the whole numbers and their opposites, which operation is now closed?

Discussion

In the calculator golf game, we used the expression *int*(*rand* x 8) + 1 to play each hole. This expression uses two functions, *int* and *rand*, that are internal to the calculator. The *rand* function outputs decimal values between zero and one. The *int* function, when applied to a positive number, outputs the whole number portion of the number. Multiplying *rand* by eight returns a decimal between 0 and 8. Applying *int* returns a whole number between 0 and 7 inclusive. Adding one results in a score between 1 and 8 inclusive. Thus the expression will always output a golf score between 1 and 8 inclusive. You then compare this score to par and record the signed number that measures your score from par.

The calculator golf game provides an opportunity to deal with negative numbers. In fact, the smaller the number, the better the score. The score on a given hole is measured from par. In this case, par acts as an arbitrary zero. It is arbitrary in the sense that the zero changes from hole to hole. It is common to use arbitrary zeros in mathematics. If we wish to study the increase in population of a city since 1985, we might make the beginning year, 1985, the arbitrary zero. Then 1990 would be interpreted as +5 and 1980 would be interpreted as –5.

Let's discuss the new numbers—negative numbers— that arose while playing calculator golf. In Chapters 1 and 2 we studied the whole numbers. We saw that the whole numbers were not closed under both subtraction and division. By including negative numbers such as those that arose in the golf game, we gain closure under subtraction. We still do not have closure under division. Negative numbers have only been accepted as useful in the last 200 years. Renaissance mathematicians referred to negative numbers as "fictitious". Leonhard Euler (1707–1783), a Swiss mathematician and physicist, actually thought that negative numbers were larger than ***infinity***. When I look at my checkbook balance, I know that negative numbers are a practical necessity and that they are not larger than infinity (Don't I wish!).

The ***opposite***, or ***additive inverse***, of a whole number is a negative number. For example, the opposite of 2 is negative 2, written –2. We need to expand our number system to include negative numbers. The collection of whole numbers and their opposites is called the set of ***integers***. If we use **I** to represent the set of integers then

$$\mathbf{I} = \{\ldots, -3, -2, -1, 0, 1, 2, 3, \ldots\}$$

There are two characteristics that are true about a number and its opposite.

1. An integer and its opposite must add to zero. Zero is called the ***additive identity***.

2. An integer and its opposite are equal distances from zero on a number line, but in opposite directions.

Investigation

10. Illustrate characteristic 1 using the integer

    a. 5

    b. 0

    c. –2

11. Draw a number line and label the integers. How is this graph different from the graph of the whole numbers?

12. Illustrate characteristic 2 using the integer

    a. 5

    b. 0

    c. –2

13. a. Select a pair of holes from Table 2 (page 153) where you were under par. Add the two negative values. Check your result on the calculator. Interpret the result in terms of the golf game.

    b. Add four other pairs of negative numbers. Record the numbers and the sums.

    c. Describe in words how to add two negative numbers.

    d. Does your rule work for the sum of –5 and –5?

14. a. Select a pair of holes from Table 2 (page 153) where one hole was under par and the other hole was over par. Add the negative and positive values. Check your result on the calculator. Interpret the result in terms of the golf game.

b. Add four pairs of numbers, one positive and one negative. Record the numbers you chose and the sums.

c. Describe in words how to add a positive number and a negative number.

d. Does your rule work for –3 + 5 and for –5 + 3? Revise your rule as necessary.

Discussion

If your scores on the first two holes were –2 and –1, the sum is –3. This indicates that you are three under par after two holes.

If your scores on the next two holes were –1 and 3, the sum is 2. This indicates you are 2 over par for those two holes.

If your scores on holes 5 and 6 were –3 and 2 respectively, the sum is –1. This indicates that you are one under par for those two holes.

The above are examples of addition of signed numbers. We can model the rules for addition by thinking of electrons (negatively–charged) and protons (positively–charged).

If we have three electrons, we would show this as ⊖ ⊖ ⊖ .

Five protons would look like ⊕⊕⊕⊕⊕ .

It might help to remember the fact from science that a proton and an electron offset each others' charge–just like the integer addition 1 + (–1) = 0.

A model of the total charge on 3 electrons plus 5 electrons is shown in Figure 1.

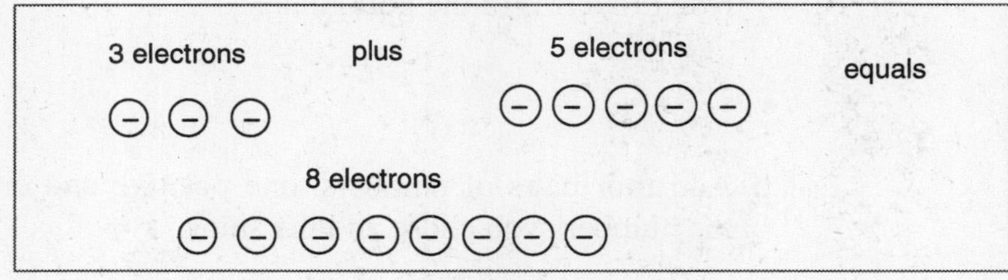

Figure 1

Algebraically we can write: $-3 + -5 = -8$.

Study the model for the addition problem $-3 + 5$. The total charge on 3 electrons plus 5 protons is shown in Figure 2.

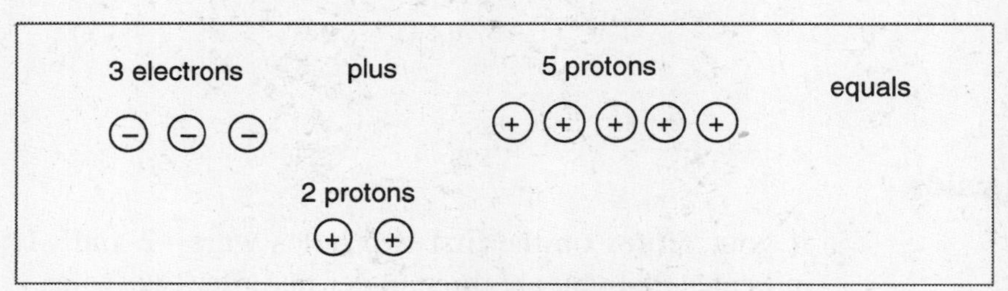

Figure 2

Another way to think of addition visually is as the result of two consecutive trips on a number line. The second part of the trip begins at the same point where the first part of the trip ends.

- A **positive** number represents movement to the **right**.
- A **negative** number represents movement to the **left**.
- The starting point is zero.

For example:

5 + 3 represents movement of 5 units right followed by movement of 3 units right ending at 8 (Figure 3).

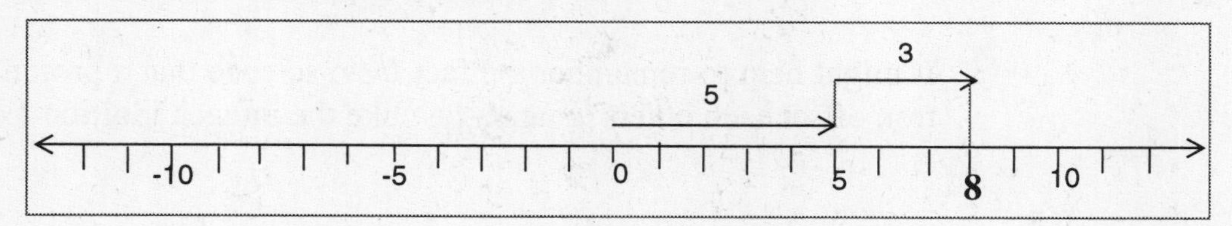

Figure 3

5 + (–3) represents movement of 5 units right followed by movement of 3 units left ending at 2 (Figure 4).

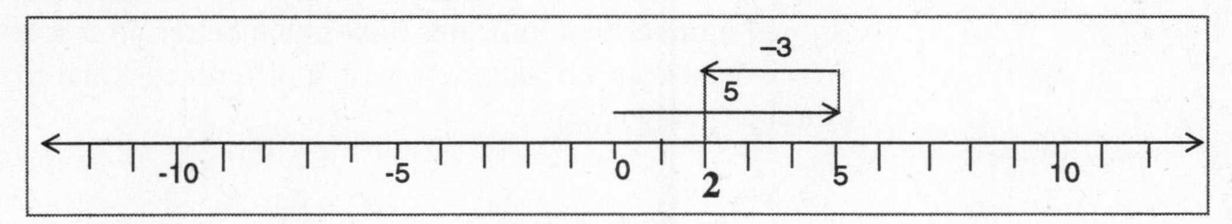

Figure 4

Be precise when you draw arrows indicating the trips.

(–5) + 3 represents movement of 5 units left followed by movement of 3 units right ending at –2 (Figure 5).

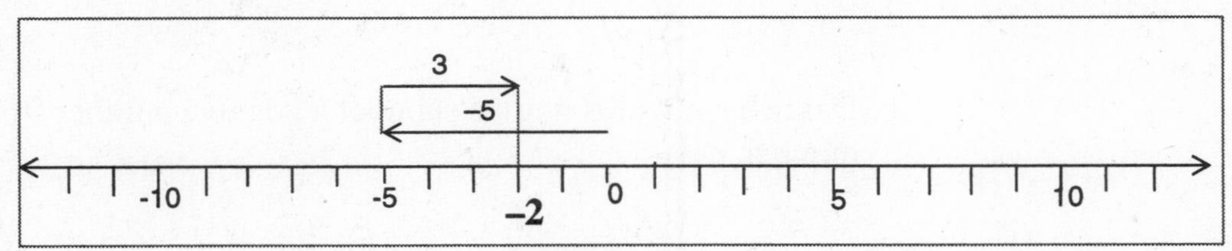

Figure 5

(–5) + (–3) represents movement of 5 units left followed by movement of 3 units left ending at –8 (Figure 6).

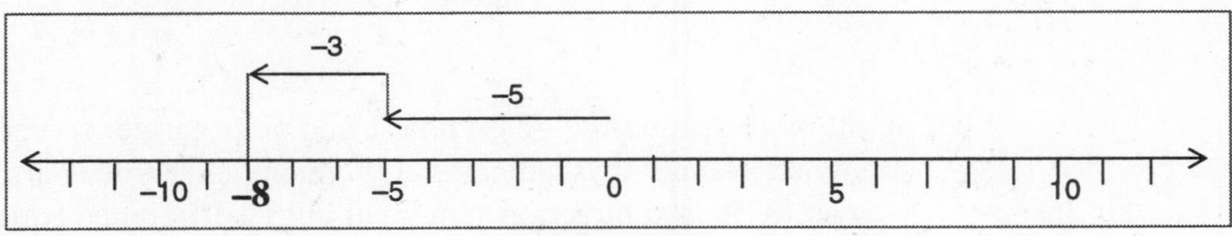

Figure 6

Next we compare scores. One way is to compare by subtraction. The quantity that is being compared to is the quantity you subtract. If I wish to compare my score to my friend's score, then my friend's score is subtracted from mine.

For example, if I scored 3 and my friend scored 1, then the subtraction statement $3 - 1 = 2$ says that my score was two strokes worse than my friend's score. Remember, negative scores are better than positive scores.

If I wish to compare my friend's score to my score, then my score is subtracted from my friend's score.

For example, if I scored 3 and my friend scored 1, then the subtraction statement $1 - 3 = -2$ says my friend's score was two strokes better than my score. The negative indicates my friend did better.

Investigation

15. a. Suppose you scored –2 on hole 1 and your partner scored 3. Write a signed number that indicates how much better your score was. Write a subtraction statement with a difference equal to the signed number you wrote.

    b. Use your calculator to subtract a positive number from a negative number. Do this for four pairs of numbers. Record the numbers you chose and the answers below.

    c. Describe in words how to subtract a positive number from a negative number.

    d. Does your rule work for (–5) – (8) and for (–8) – (5)? Revise your rule as necessary.

16. a. Suppose you scored –2 on hole 1 and your partner scored 3. Write a signed number that indicates how much worse your partner's score was. Write a subtraction statement with a difference equal to the signed number you wrote.

    b. Use your calculator to subtract a negative number from a positive number. Do this for four pairs of numbers. Record the numbers you chose and the answers below.

    c. Describe in words how to subtract a negative number from a positive number.

    d. Does your rule work for (5) – (–8) and for (8) – (–5)? Revise your rule as necessary.

17. a. Suppose you scored –3 on hole 5 and your partner scored –1. Write a signed number that indicates how much better your score was. Write a subtraction statement with a difference equal to the signed number you wrote.

b. Suppose you scored –3 on hole 5 and your partner scored –1. Write a signed number that indicates how much worse your partner's score was. Write a subtraction statement with a difference equal to the signed number you wrote.

c. Use your calculator to subtract four pairs of negative numbers. Record the numbers you chose and the answers below.

d. Describe in words how to subtract two negative numbers.

e. Does your rule work for $(-5) - (-8)$ and for $(-8) - (-5)$? Revise your rule as necessary.

f. Does your rule work if you subtract –5 from –5? Revise your rule as necessary.

18. a. Enter -5^2 on your calculator. Record the result.

b. Enter $(-5)^2$ on your calculator. Record the result.

c. Describe the difference between parts a. and b. Specifically address the order of operations.

Discussion

Subtraction is required to calculate your scores relative to your partner's score. If you scored –2 on hole 1 and your partner scored 3, then your score relative to your partner's is –5. The subtraction statement is $(-2) - 3 = -5$.

On the other hand, your partner's score relative to yours is 5. The subtraction statement is $3 - (-2) = 5$.

If you scored –3 on hole 5 and your partner scored –1, then your score relative to your partner's is –2. The subtraction statement is $(-3) - (-1) = -2$.

Your partner's score relative to yours is 2. The subtraction statement is $(-1) - (-3) = 2$.

To help understand subtraction, we need to investigate the idea of opposite further.

Finding an opposite of a number means to change the sign of the number. Thus the opposite of a positive number is a negative number and the opposite of negative number is a positive number.

If a represents a number, the opposite of a is represented by $-a$.

A function machine for the operation of ***oppositing*** appears in Figure 7.

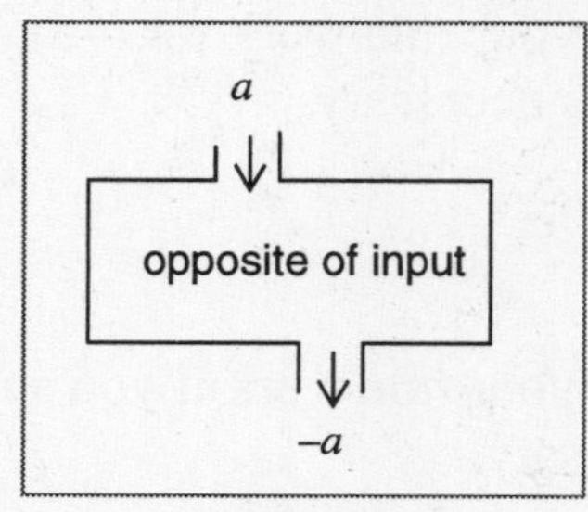

Figure 7

For example, if $a = 17$
then $-a = -(17) = -17$ (Figure 8).

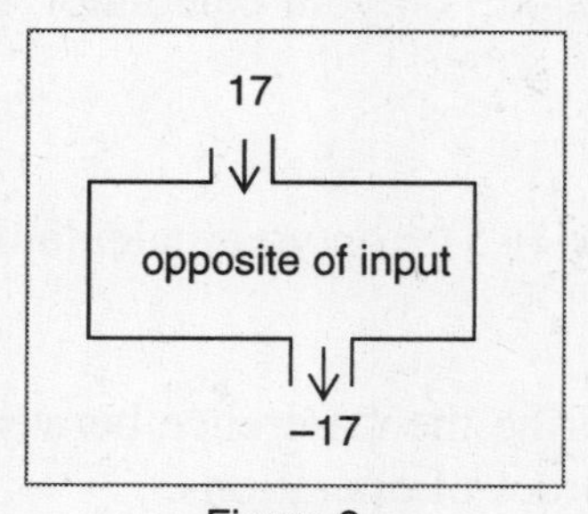

Figure 8

–(17) = –17 is read "the opposite of 17 is negative 17". Notice that if the input is positive, the output is negative.

Also, if $a = -32$
then $-a = -(-32) = 32$ (Figure 9).

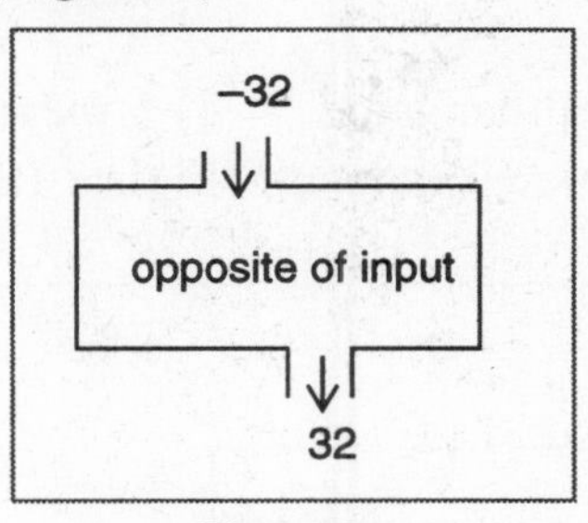

Figure 9

– (–32) = 32 is read "the opposite of negative 32 is 32". Notice that if the input is negative. the output is positive.

The operation of finding an opposite, symbolized by "–", is a ***unary operator*** (unary since it operates on only one number at a time). Be careful! This symbol (–) is used in three different ways:

- To indicate the unary operation of oppositing such as –(5), read "the opposite of 5".
- To indicate the ***binary operation*** (binary since it operates on two numbers at a time) of subtraction such as 5 – 7, read "5 minus 7".
- To indicate a negative number such as –5, read "negative 5".

There is an important distinction when using the symbol "–".

- The symbol "–" placed in front of a number is read "negative of the number". We read –5 as negative five.
- The symbol "–" placed in front of a variables is read the "opposite of the variable". We read $-a$ as the opposite of a.

When finding an opposite, we operate on one number to obtain a second number.

When doing a computation involving several operations, we need to recall order of operations. First remember that

unary operations take precedence over binary operations.

We now have two unary operations: exponentiation and oppositing.

Exponentiation takes precedence over oppositing in the absence of grouping symbols.

So -5^2 means that we must first square 5 and then take the opposite of the answer (Figure 10a).

On the other hand $(-5)^2$ means that we take the opposite of 5 first and then square the result (Figure 10b).

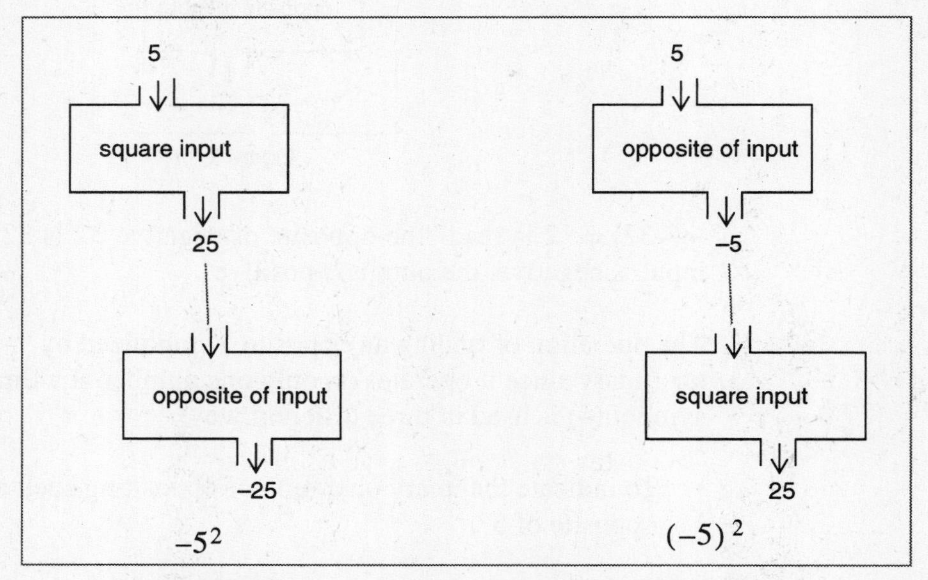

Figure 10a Figure 10b

Investigation

19. a. If $t = 7$ then $-t =$ _______.

    b. If $b = -2$ then $-b =$ _______.

    c. If $r = 823$ then $-r =$ _______.

20. Do each of the following computations. Compare the results of each subtraction and the addition problem.

    a. $7 - 3 =$ _____

    $7 + (-3) =$ _____

    b. $7 - (-3) =$ _____

    $7 + 3 =$ _____

    c. $-7 - 3 =$ _____

    $-7 + (-3) =$ _____

    d. $-7 - (-3) =$ _____

    $-7 + 3 =$ _____

21. a. If you understand that subtraction reverses direction, draw the subtraction 7 – 3 on a number line.

    b. Draw the addition 7 + (–3) on a number line.

    c. Compare the results of parts a. and b.

22. a. Rewrite (–2) – (5) as an addition.

    b. Rewrite –4 – (–3) as an addition.

    c. Rewrite 5 – (–4) as an addition.

    d. Describe how to rewrite a subtraction as an addition.

Discussion

From Investigations 20–22, we see that a subtraction ***is equivalent to*** adding the opposite of the ***subtrahend***. To see why, let's look at subtraction and addition on a number line.

Figure 11 displays the subtraction 7 – 3.

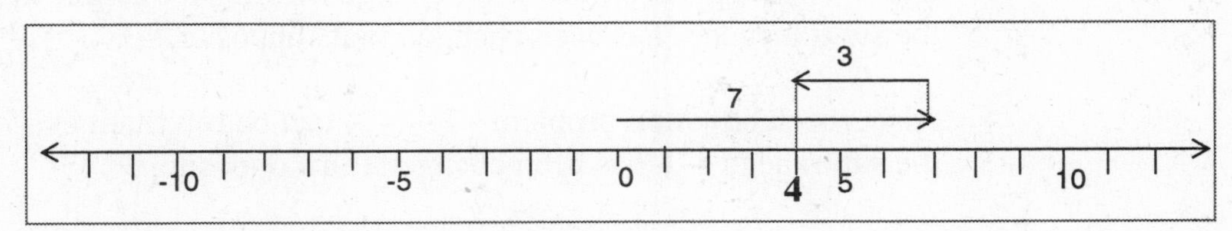

Figure 11

Figure 11 displays the subtraction 7 – 3.

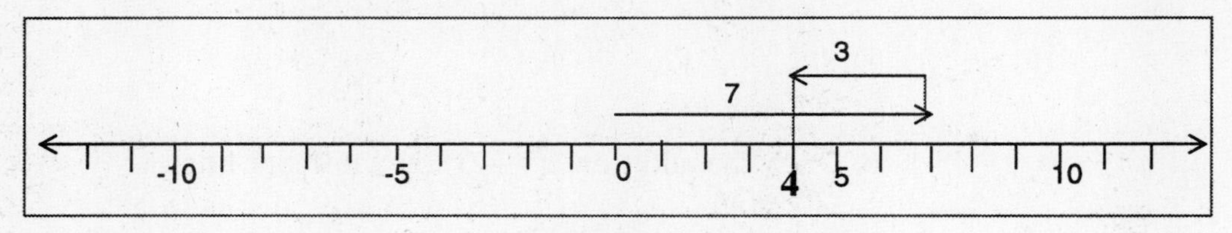

Figure 11

Figure 12 displays the addition 7 + (–3).

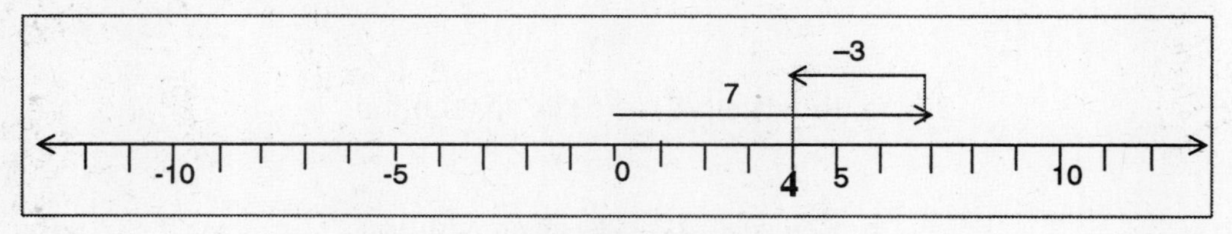

Figure 12

Converting subtraction to addition is accomplished by making two changes.

- Change the operation from subtraction to addition and
- Change the number being subtracted (subtrahend) to its opposite (Figure 13).

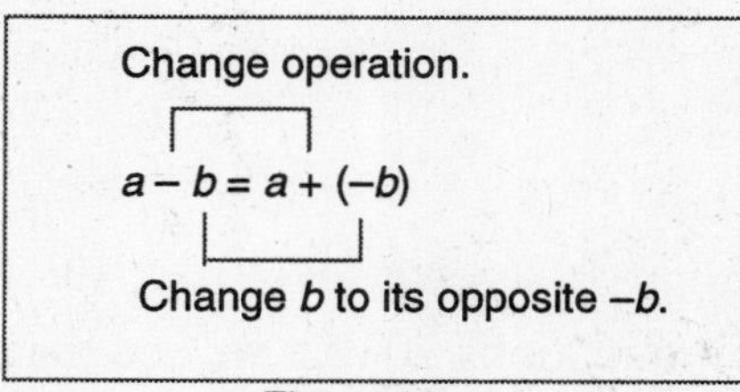

Figure 13

For example, the subtraction problem –7 – 3 is read "negative seven subtract 3". This can be rewritten as –7 + (–3) read "negative seven plus negative 3". The operation has been changed from subtraction to addition and the subtrahend, 3, has been changed to its opposite, –3.

Also, the subtraction problem –7 – (–3) can be rewritten as –7 + (3) where the subtrahend, –3, has been changed to its opposite, 3.

Once the subtraction has been rewritten as an addition, we can use the electron–proton approach or the trip followed by a trip approach to find the answer.

Explorations

1. List and define words in this section that appear in ***italics bold*** type using your own words.

2. Find the opposite of each of the following:

   a. 6 b. –5
   c. 0 d. $5x$
   e. $-3y$ f. $x + y$

3. Use the electron–proton model to demonstrate the following additions.

   a. 2 + (–5) b. (–7) + (–3)
   c. 9 + (–2)

4. Use the number line model to demonstrate the following additions.

   a. 8 + (–2) b. 6 + (–9)
   c. (–4) + (–3)

5. Write each of the following as an addition. Use either the electron–proton or the number line model to demonstrate the resulting addition.

   a. 3 – (7) b. –2 – (6)
   c. –4 – (–5) d. –9 – (–7)

6. Refer to Investigation 15d (page 160).

   a. Is this an example of commutativity of subtraction?
   b. Write an example that shows subtraction is not commutative.

7. Use your calculator to multiply five pairs of negative numbers.

   a. Record the numbers you chose and the answers.
   b. Describe in words how to multiply two negative numbers.

8. Use your calculator to multiply five pairs of numbers, one positive and one negative.

   a. Record the numbers you chose and the answers.

   b. Describe in words how to multiply a positive and a negative number.

9. Use your calculator to divide five pairs of negative numbers.

   a. Record the numbers you chose and the answers below.

   b. Describe in words how to divide two negative numbers.

10. Use your calculator to divide five pairs of numbers, one positive and one negative.

    a. Record the numbers you chose and the answers below.

    b. Describe in words how to divide a positive and a negative number.

11. Evaluate.

    a. $-8 - 7$
    b. $(-8)(-7)$
    c. $-(-35)$
    d. $-15 \div 3$
    e. $-13 + 2$
    f. $6 + (-9) - 5 - (-7)$
    g. $(-3)(8) + (-2)(-6)$
    h. $(8 + (-2))(5 - 9)$
    i. $17 - (-9) + (-54) - 38$
    j. $-24 \div (-6)(-3)$
    k. $4(-3)^2 - 2(2)^3 + (-3)(5)$

12. Given $T(r) = 5r - 2$. Find $T(-3)$.

13. Given $y(x) = 3x^2 - x - 4$. Find $y(-5)$.

14. We used the expression *int*(*rand* x 8) + 1 to generate golf scores between 1 and 8 inclusive.

    a. Write an expression that would generate golf scores between 1 and 10 inclusive.

    b. Write an expression that would randomly generate numbers between –3 and 3 inclusive.

This type of function requires a new type of function machine. The input must be evaluated to determine which process to use. Upon entering the function machine, a decision about the input is made. Based on the decision, we follow *exactly one* of the multiple paths. Figure 1 illustrates this type of function machine for the given problem situation.

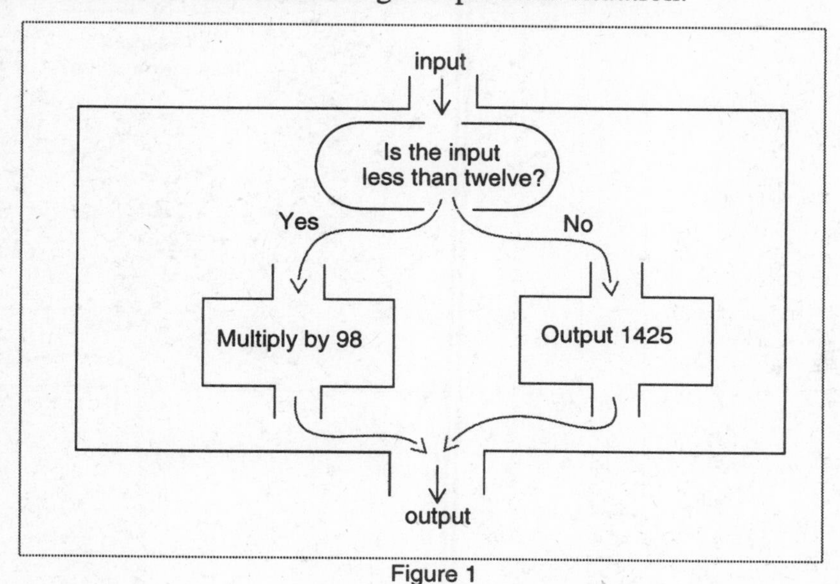

Figure 1

The oval at the beginning of the function machine indicates a decision must be made first. We follow only one path based on the decision. As a result, there is only one path to the output for a given input.

Figure 2 illustrates the machine when the input is ten.

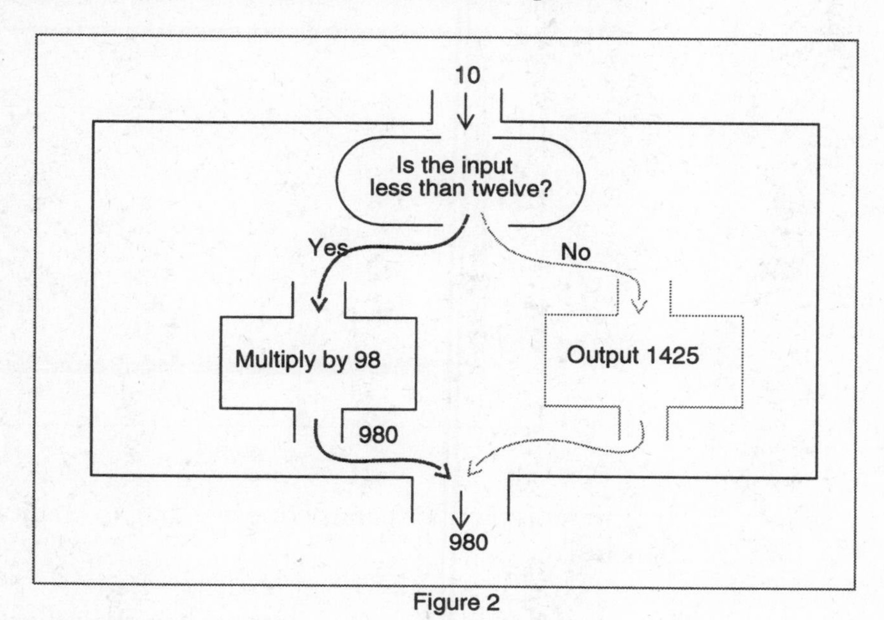

Figure 2

Figure 3 illustrates the machine when the input is fifteen.

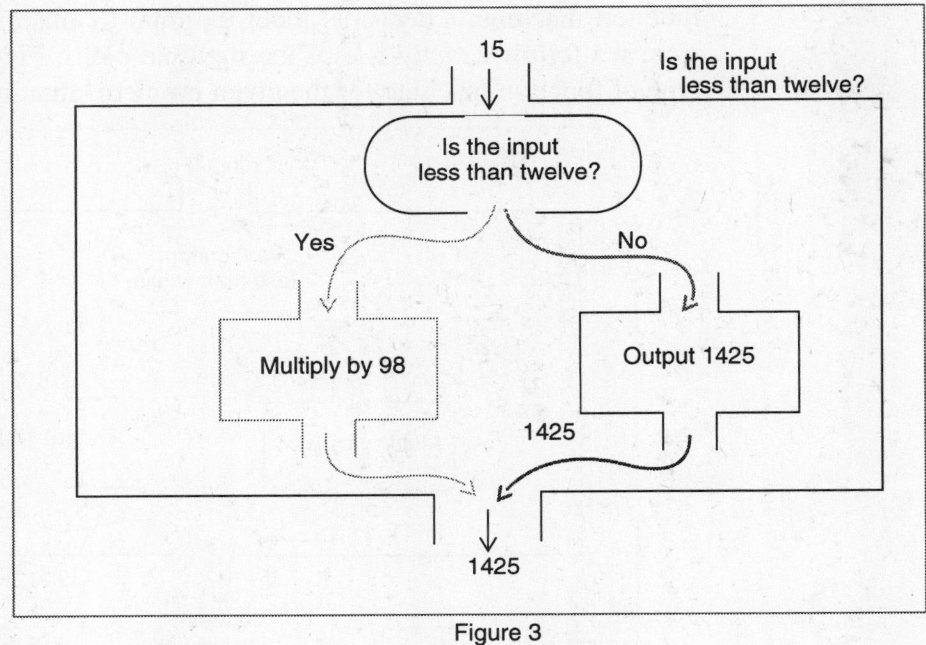

Figure 3

Finally, let's look at the graph of this relationship (Figure 4).

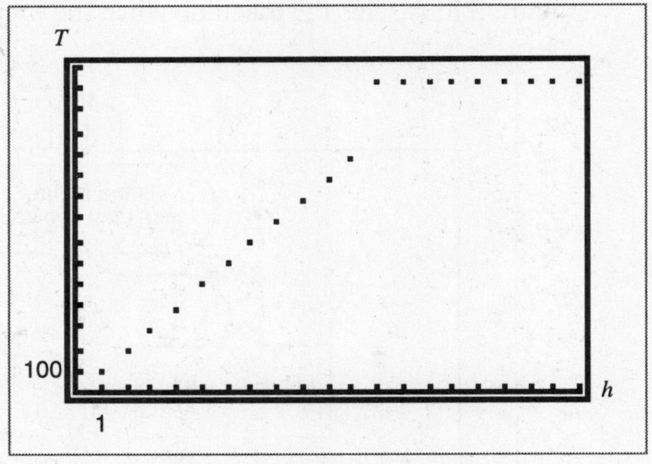

Figure 4

The graph displays all points corresponding to credit hours between 1 and 20 inclusive.

Next we investigate a very important mathematical relationship that is another example of a piecewise defined function. We begin with a problem situation involving the stock market.

Investigation

Problem situation: Ivan B. was a stockbroker who used a simple test to determine whether to buy or sell a stock that was of interest. If the change in the stock from opening to closing exceeded two points, he sold the stock if he owned it and bought the stock if he did not.

5. Use the price of the following stocks to decide which action (buy, sell, or wait) is determined by the rule. Ivan B. owns the stocks indicated by the *.

Stock	Open	Close	Change Δx	Action
* IBM	70	72		
AMOCO	53	49		
Apple	17	24		
* Casio	20	18.5		
Sharp	13	13		
*TI	19		–3	
HP	43	40.75		
* SPC		11	7	
DMW	62		3	
Motorola		59	–2	

Table 2: Stock Market

Discussion

When computing an amount of change, the question of what order to subtract is an important one. Since we are interested in change, we want to know the amount or ***magnitude*** of change, as well as the ***direction*** of the change. If the quantity gets larger, the change should be positive, but if the quantity gets smaller, the change should be negative. This can be accomplished by subtracting the ***initial value*** of the quantity from its ***final value.***

If p represents the price of a stock, then

p_1 represents the initial value,

p_2 represents the final value, and

Δp denotes the change in p.

The change in p is the final value minus the initial value, or

$$\Delta p = p_2 - p_1 .$$

The Greek letter ***delta***, represented by the symbol Δ, is used in mathematics for the phrase "change in". So the symbol Δp means "the change in p". Notice that we use small numbers below and to the right of the variable, such as the 1 in p_1. These are called ***subscripts***. Mathematicians do this when they use the same variable in several different settings. We are using the variable p to represent both the initial and final price. The subscripts 1 and 2 on p_1 and p_2 represent a way symbolically to distinguish between the two values. When you see p_1 and p_2, read these as "p sub 1" and "p sub 2". The word "sub" indicates subscript.

In this case, the subscript 1 is used to designate the initial value and the subscript 2 is used to designate the final value.

Investigation

6. If x represents the money in my petty cash fund, what notation would be used to designate the

   a. amount of money at the beginning of the day (initial value)?

   b. amount of money at the end of the day (final value)?

   c. the change in cash during the day?

7. Refer to Investigation 6.

   a. What would a negative change imply?

   b. What would a positive change imply?

8. Complete Table 3 by finding Δx. Assume that x_1 represents the initial value and x_2 represents the final value.

Initial value x_1 in \$	Final value x_2 in \$	Change in petty cash Δx in \$
5	3	
3	5	
3	3	
0	7	
7	0	
0	-2	
–2	0	
–3	4	
4	–3	
1		–2
	–3	3
t		0
t		4
	t	–4
–5		0
		0
		5
		–5

Table 3: Daily Change in Petty Cash Account

9. Determine the sign of Δx if:

a. $x_1 = x_2$

b. $x_1 < x_2$

c. $x_1 > x_2$

10. a. Last week, the DMW stock had the opening prices listed below. Complete Table 4 showing the daily changes. Assume the close on Monday is the opening price on Tuesday.

Day	Opening	Close	Change
Monday	17		
Tuesday	21		
Wednesday	19		
Thursday	4		
Friday	7	8	

Table 4: A week at the stock market

b. Write a mathematical statement using Δp to show other possible changes.

11. The temperature at noon yesterday was 27° C. The temperature dropped three degrees each hour all day.

a. Find the temperature at 4:00 p.m. and the change in temperature Δt.

b. Find the temperature at 9:00 a.m. and the change in temperature Δt.

12. Ivan B.'s rule for when to buy or sell a stock required that all he needed to know was the magnitude of the change. He simply looked at the problem and changed the sign of Δx if it was negative. Use your results from Investigation 5 to determine which Δx's change.

Discussion

What we need for Ivan's rule is a function that makes any quantity positive. We don't want to change the magnitude (size) of the quantity, we only want to make certain it is positive. Fortunately for us, there are many situations where it is convenient to talk about a quantity as though it were positive, so mathematicians have already invented this function. (Don't they think of everything?) The function is called the ***absolute value function***. The notation used to indicate the absolute value of x is $|x|$.

Since we want to change the sign of a quantity *only when it is negative*, we must be careful how we define this function. We need a piecewise defined function whose definition focuses on the sign of the input, not the size of the input. If the input is a positive or zero quantity initially, we don't want to change anything; but if the input is negative, then we need to take the opposite of the input so that our final result is positive.

Investigation

13. a. Complete Table 5 where the input is x and the output is the absolute value of x, written $|x|$.

x	$\lvert x \rvert$
5	
–7	
0	
–3	
8	
2	
–17	

Table 5: Absolute value 1

b. For each answer in part a, how does the output (answer) compare to the input (the original value of x)?

14. Write a definition of $|x|$ as a piecewise defined function.

15. Compare your definition with others in your group. Refine it into one you all agree with. Use that definition to find the absolute value of the following values for x, explaining what action you took.

x	$\lvert x \rvert$	Action
5		
−2		
5 + 3		
5 − 3		
3 − 5		
−3 + 1		
−3 − 5		
−3 − (−5)		

Table 6: Absolute Value 2

16. Suppose that x is some number greater than ten for each of the following quantities. Find the absolute value of each. That means, write each quantity without absolute value signs, so that you know that quantity is positive. Explain how you know the action is positive.

a. $\lvert x \rvert$

b. $\lvert x + 5 \rvert$

c. $\lvert 5 - x \rvert$

d. $\lvert 2x \rvert$

e. $\lvert 27 - 3x \rvert$

Discussion

As you noticed in the preceding investigations, unless you know where a quantity is positive and where it is negative, you can't tell what the absolute value of a quantity is. Absolute value is an unary operation that has one input and one output. Before the output is determined we must know the sign of the input. If the input is positive or zero then the output is the same as the input. However if the input is negative then the output is the opposite of the input. The function machine appears in Figure 5.

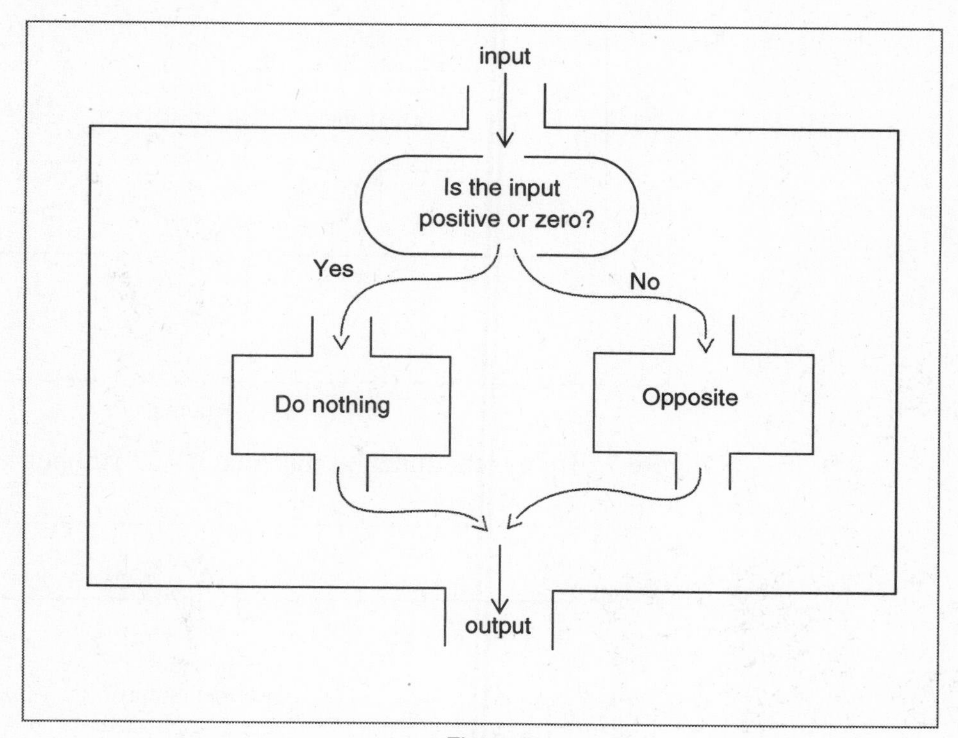

Figure 5

Notice the decision oval at the beginning of the function. Based on the answer to the decision, we either do nothing if the answer is yes or find the opposite of the input if the answer is no. Just like the piecewise defined function for the tuition charges, we follow only one path for each input. Thus there is exactly one output for a given input.

Let's look at the process for two numerical inputs first.

Figure 6 displays the absolute value function machine with input 43.

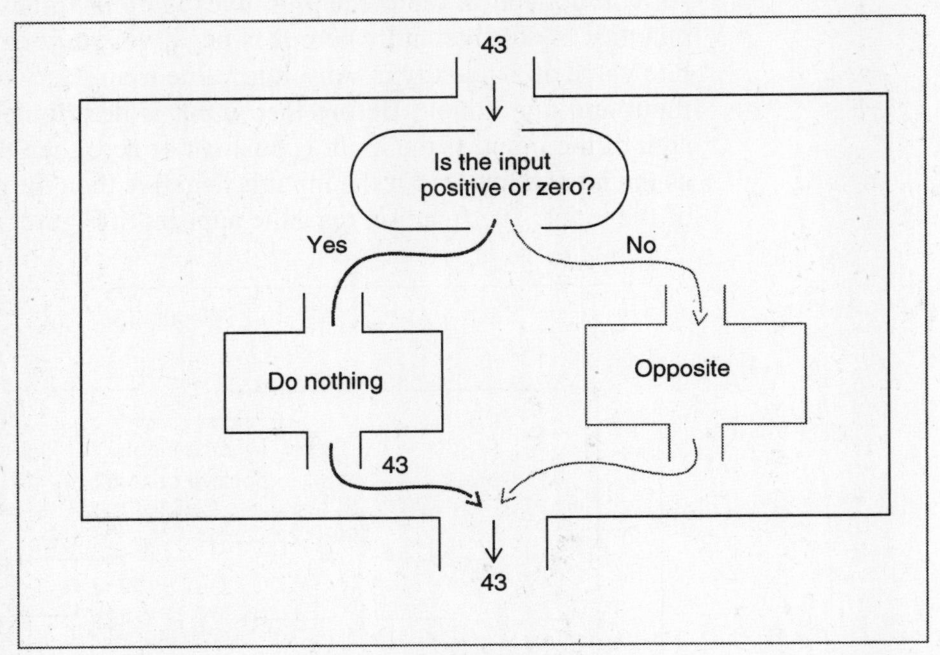

Figure 6

Figure 7 displays the function machine if –27 is input.

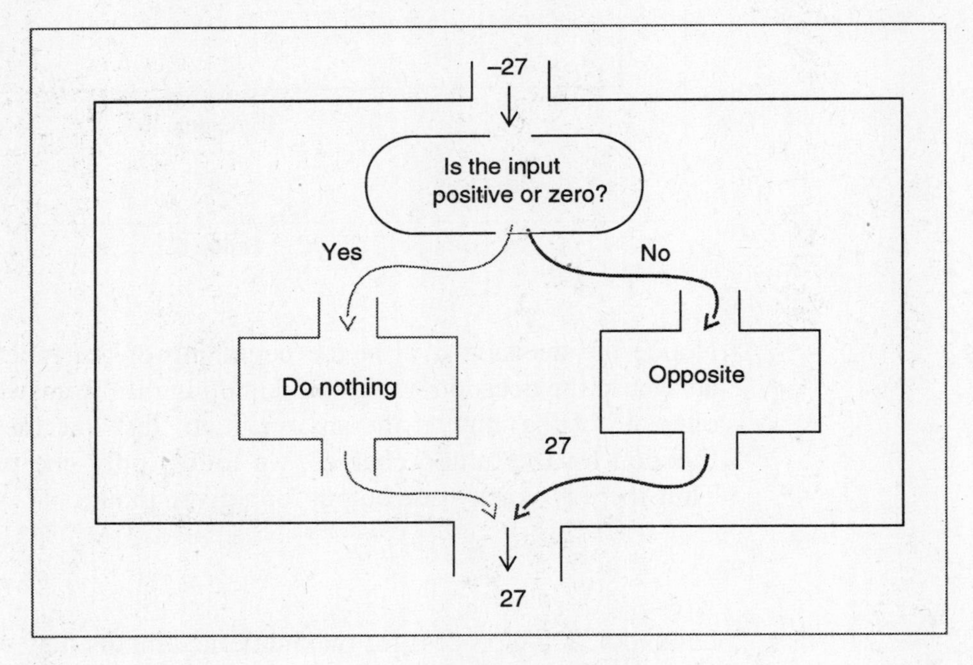

Figure 7

Let's look at the process for two of the problems in Investigation 16.

Given $|x+5|$. The input is $x+5$ and x is some number greater than 10. For any value of x larger than 10, the input $x+5$ is positive. Figure 8 displays the function machine for this problem.

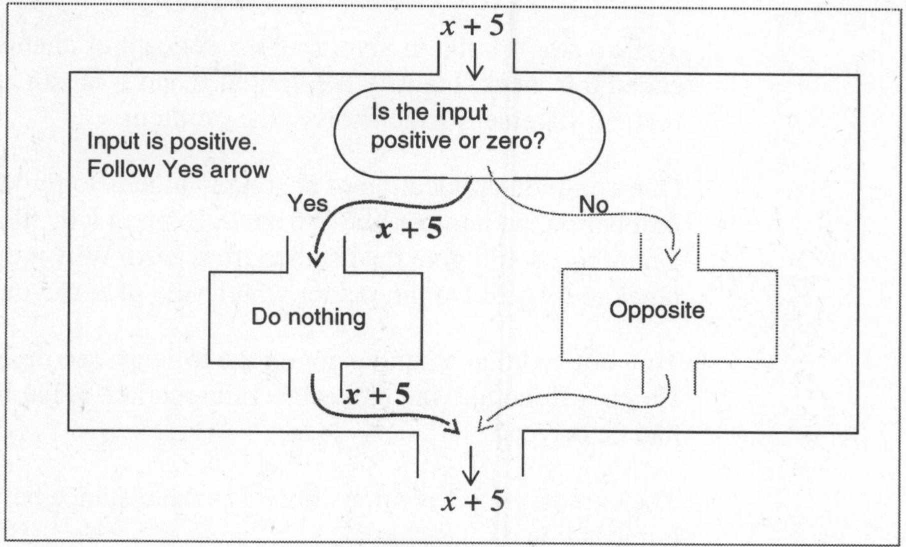

Figure 8

We see that

$$|x+5| = x+5 \text{ when } x \text{ is larger than } 10.$$

Given $|5-x|$. The input is $5-x$ and x is some number greater than 10. For any value of x greater than 10, $5-x$ is negative. Figure 9 displays the function machine for this problem.

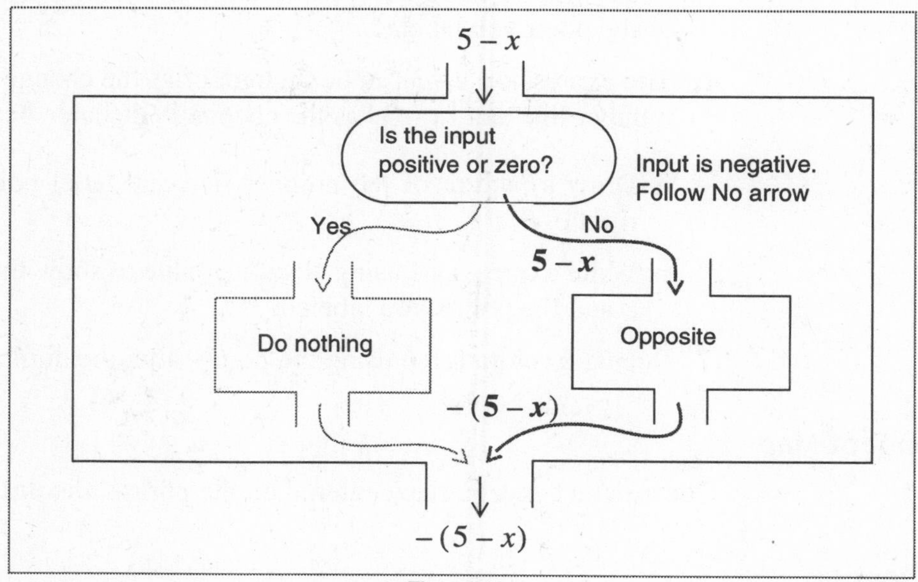

Figure 9

We see that

$$|5-x| = -(5-x) \text{ when } x \text{ is larger than } 10.$$

Explorations

1. List and define words in this section that appear in ***italics bold*** type using your own words.

2. Write a story problem involving the concept of change that you experienced this week. Identify the variables and write an appropriate mathematical statement which solves the problem.

3. One common application of absolute value is to find the distance a number on the number line is from 0. Explain why the absolute value of a number would give its distance from zero. Why would you want distance to be positive, no matter which side of zero you were on?

4. Tom noticed that when he got on the tollway, the mile marker was 143. He got off the tollway at the 109 mile marker. What is Δm? How far had he driven?

5. The expression $|x|$ is often defined as the distance between x and 0 on a number line.

   a. Interpret $|6|$ and $|-43|$ in terms of this definition.

   b. Draw a number line and label a point to the right of zero. Write an expression using absolute value to show the distance between 0 and the point you labeled.

   c. Draw a number line and label a point to the left of zero. Write an expression using absolute value to show the distance between 0 and the point you labeled.

6. The expression $x - 3$ can be thought of as the change from 3 to x on the number line and $|x - 3|$ as the change in distance between 3 and x.

   a. Draw a picture of the number line and label points to the left and right of 3.

   b. Write expressions using absolute value to show the distance between 3 and the points you labeled.

7. Repeat Exploration 6 using the point -4 on the number line.

Concept Map

Construct a concept map centered on the phrase **absolute value**.

Reflection

Your cousin is having a hard time understanding the function concept. Write a paragraph using function machines to help her understand the idea of a function.

Section 4.3 Graphing with Integers

Purpose

Extend graphing in two dimensions using the integers.

Discussion

Recall the Mayan Mixup discussed earlier in the text. You might want to review the rules of the game (Section 3.4). We will modify the Reference Map so that the vertex (0, 0) is in the middle of the grid rather than at the lower left corner (Figure 1). The Search Plan (Table 1) is changed accordingly.

Investigation

1. Begin the game by creating a reference map of your four sites. Draw your treasure sites.

2. Record below the vertices where each site is located.

3. On your Search Plan, select and record the vertices where you think the opponent's sites are located. Note the vertices you locate.

4. As your opponent names the vertices where he/she thinks your sites are located, record those vertices on your Reference Map.

5. Complete Round 1 by trying to locate the sites of your opponent.

6. Complete Round 1.

   a. Record the vertices you selected for your attempts to locate your opponent's treasures.

   b. Record the vertices where you scored a hit.

   c. Record the vertices selected by your opponent.

   d. Record the vertices where you took a hit.

7. Complete the game. Keep track of all attempts and successful probes in each round, using the Search Plan and Reference Map on the next page.

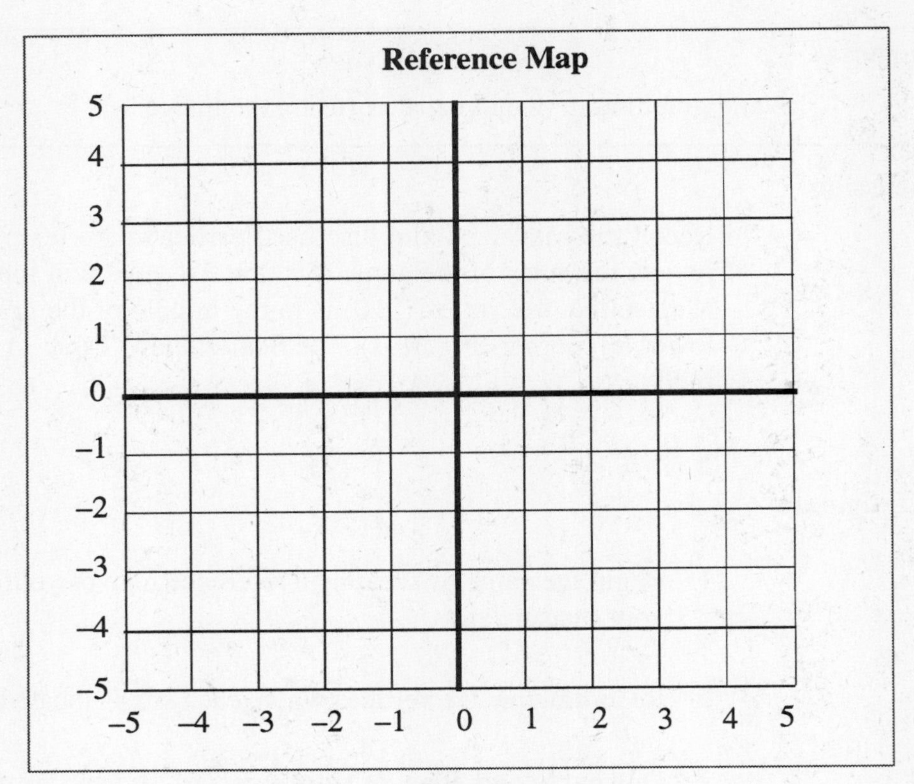

Figure 1

5											
4											
3											
2											
1											
0											
–1											
–2											
–3											
–4											
–5											
	–5	–4	–3	–2	–1	0	1	2	3	4	5

Table 1: Search Plan

Discussion

When we graph with negative numbers, we extend the horizontal and vertical axes so that they represent number lines containing all the integers.

On the horizontal axis, the input axis, numbers increase as we move from left to right.

On the vertical axis, the output axis, numbers increase as we move up.

Let's look at a typical placement of sites on the Reference Map (Figure 2).

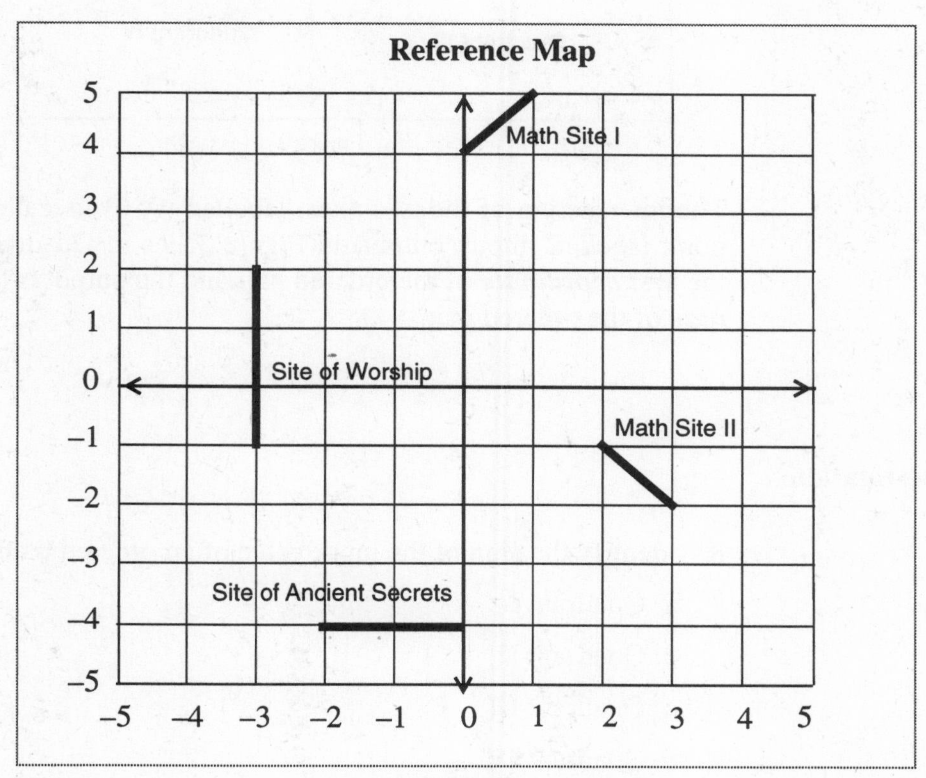

Figure 2

We have the following site locations.

- Site of Worship: (–3, –1), (–3, 0), (–3, 1), (–3, 2). .
- Site of Ancient Secrets: (–2, –4), (–1, –4), (0, –4).
- Mathematical Writings Site I: (0, 4), (1, 5).
- Mathematical Writings Site II: (2, –1), (3, –2).

Notice that the two number lines divide the plane into four parts, called ***quadrants***, which are numbered as shown in Figure 3.

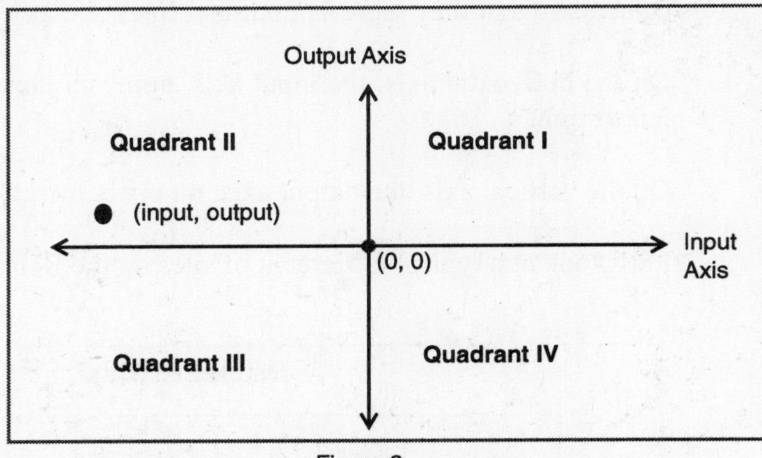

Figure 3

The intersection of the two axes, labelled (0, 0), is called the ***origin***. The point labelled (input, output) in Figure 3 lies in Quadrant II. The input is the ***first coordinate*** of the ordered pair and the output is the ***second coordinate*** of the ordered pair.

Investigation

8. Identify the sign of the input value of an ordered pair that graphs in
   a. Quadrant I
   b. Quadrant II
   c. Quadrant III
   d. Quadrant IV.

9. Identify the sign of the output value of an ordered pair that graphs in
   a. Quadrant I
   b. Quadrant II
   c. Quadrant III
   d. Quadrant IV.

Problem Situation: A wholesaler sells Chicago Bulls T–shirts to Okimbe for a total of $1000. Okimbe sells the shirts for $10 each. Table 2 displays the relationship between the number of shirts Okimbe sells and Okimbe's profit.

Number of shirts sold	Profit ($)
0	–1000
50	–500
100	0
150	500
200	1000
250	1500

Table 2: Okimbe's profit on T–shirt sales

10. Express the information in the Table 2 as ordered pairs.

Investigations 11–14 refer to Figure 4 where we graph the information given in Table 2.

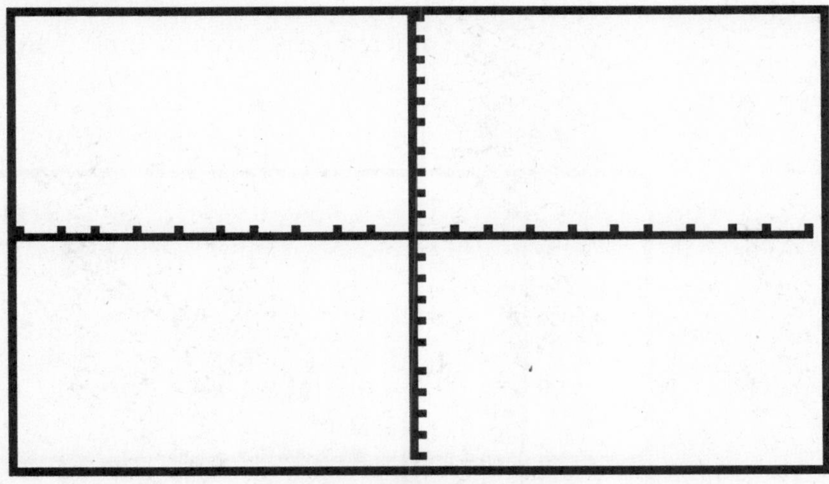

Figure 4

11. Use words to label the input ***axis*** and output ***axis***.

12. Record reasonable values below the tick marks on the horizontal axis in order to represent the input.

13. Record reasonable values to the left of the tick marks on the vertical axis to represent output.

14. Mark and label a point for each ordered pair specified by Table 2.

Discussion

When plotting points in the plane, we have the following correspondence between the quadrant location and the signs on the input and output (See Figure 5).

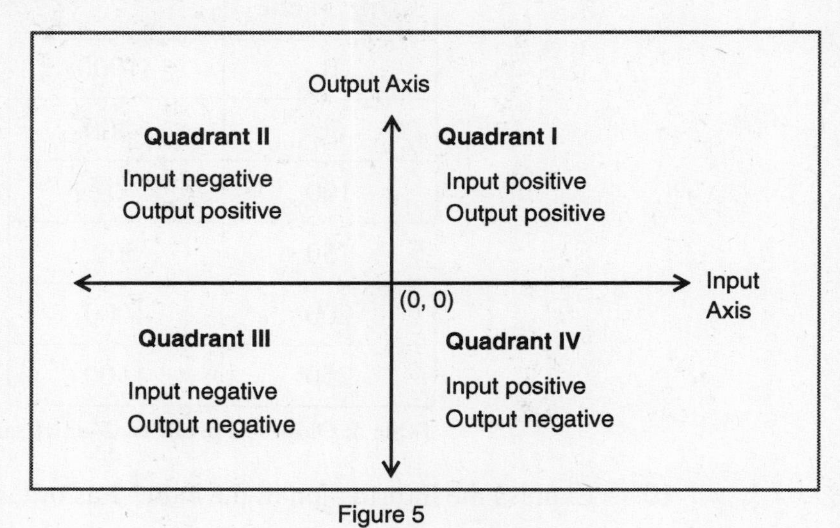

Figure 5

The graph of the ordered pairs given in Table 2 are displayed in Figure 6.

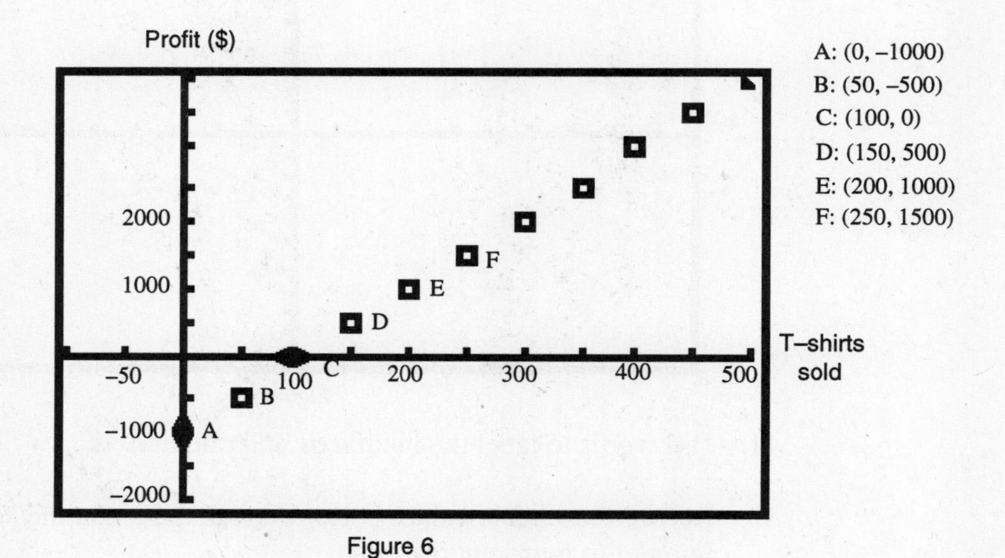

Figure 6

Explorations

1. List and define words in this section that appear in ***italics bold*** type using your own words.

2. Given the Reference Map (Figure 7), identify the locations of each site by recording the ordered pairs of vertices where each site has been located.

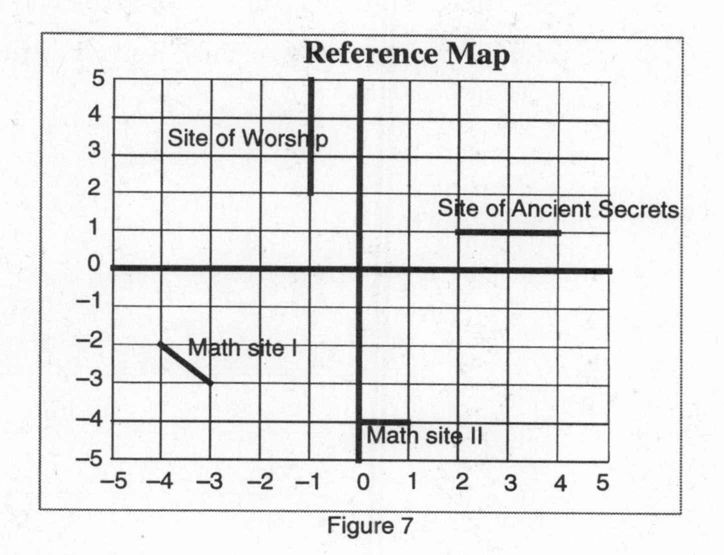

Figure 7

3. Graph the ordered pairs (–5, 40), (3, –50), (–2, –10), (6, 20), (0, –30), (–4, 0), (7, 0), and (0, 10) on the grid given in Figure 8.

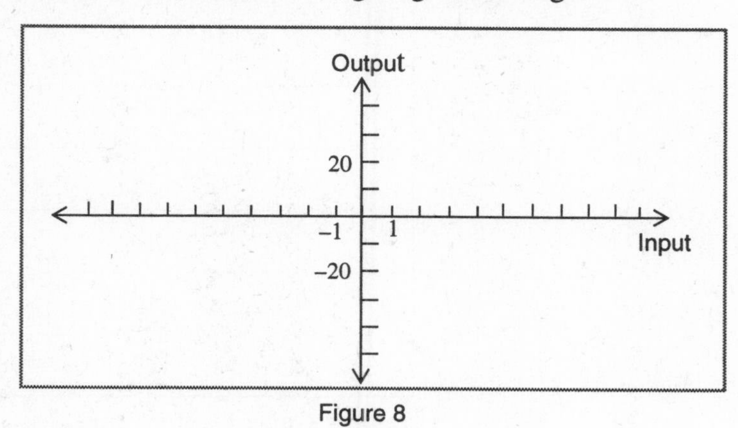

Figure 8

4. Refer to the T–shirt problem discussed in this section.

   a. Describe how to find Okimbe's profit if you know how many T–shirts were sold.

   b. Draw a function machine displaying the relationship between the number of T–shirts sold and the profit.

   c. Write an algebraic representation (equation) that displays the relationship between the number of T–shirts sold and the profit.

5. Make up your own problem similar to the T–shirt problem.

6. For your problem,

   a. construct a table of at least five input/output pairs.

   b. graph the ordered pairs in your table.

   c. draw a function machine displaying the relationship described in your problem.

7. Complete Table 3.

Integer	Opposite of integer
–5	
–4	
–3	
–2	
–1	
0	
1	
2	
3	
4	
5	

Table 3: Opposites of integers

8. Graph the relationship displayed in Table 3.

9. Complete Table 4.

Integer	Absolute value of integer
−5	
−4	
−3	
−2	
−1	
0	
1	
2	
3	
4	
5	

Table 4: Absolute value of integers

10. Graph the relationship displayed in Table 4.

11. Calculate:

a. $-5 + 17$

b. $2 - 11$

c. $13 + (-8)$

d. $-5 - 9$

e. $-7 - (-3)$

f. $|-9|$

g. $-|5|$

h. $-|-6|$

i. $(-7)(4)$

j. $(-8)(-3)$

k. $\frac{18}{-9}$

l. $-\frac{18}{9}$

m. $\frac{-18}{9}$

n. $\frac{-18}{-9}$

o. $-\frac{-18}{-9}$

p. $|-3| - 5$

q. $-4(|-8|)$

r. $-|-2| + |-9|$

12. The function $P(m) = 3m - 50$ expresses the profit P (in dollars) from the sale of costume jewelry as a function of the number m of pieces sold.

a. Complete Table 5.

Pieces sold	Profit ($)

Table 5: Profit on sale of costume jewelry

b. Describe in words how to evaluate the function.

c. Graph the data from Table 5.

d. Use numerical estimation to determine the number of pieces of jewelry that must be sold to realize a profit of $104.

Concept Map

Construct a concept map centered on the phrase **graphing ordered pairs**.

Reflection

Clearly describe the steps required to graph the ordered pair (a, b) where a and b represent integers.

Section 4.4 Using a Graphing Utility

Purpose

Introduce the use of a graphing utility to construct graphs.

Discussion

Recall the Mayan Mixup game that we have played previously. The entire game was played on a two–dimensional grid and required plotting ordered pairs to locate specific treasures. Let's now make the screen of your graphing utility the gameboard.

Investigation

Assume that treasures are located at the following vertices.

Site of worship: (–5, 3), (–6, 4), (–7, 5), (–8, 6).
Site of ancient secrets: (17, –2), (17, –3), (17, –4).
Mathematical writings 1: (–9, –11), (–10, –11).
Mathematical writings 2: (0, 6), (1, 7).

1. Record all the ordered pairs in Table 1.

First coordinate	Second coordinate

Table 1: Treasure locations

2. Enter the values of Table 1 into your calculator. If you do not know how, ask your instructor or a classmate, or counsult your manual.

3. Graph the values in Table 1 on your calculator in a graphing window so that both the x–axis and y–axis vary from –10 to 10. If you do not know how, ask your instructor or a classmate, or consult your manual.

4. a. What sites are displayed on the graph? Why?

   b. What sites are not displayed on the graph? Why?

Now let's display the location of all the treasure. This requires us to set the screen coordinates so that we can "see" all the points in the viewing window.

5. a. List the smallest and largest first coordinate value from Table 1.

   b. List the smallest and largest second coordinate value from Table 1.

6. a. Write a number smaller than the smallest first coordinate. Call this number **Xmin.**

   b. Write a number larger than the largest first coordinate. Call this number **Xmax.**

   c. Write a number smaller than the smallest second coordinate. Call this number **Ymin.**

   d. Write a number larger than the largest second coordinate. Call this number **Ymax.**

7. Use your answers to Investigation 6 to change the viewing window on your graphing utility. If you do not know how, ask your instructor or a classmate, or consult your manual. Graph the data in Table 1.

Discussion

Previously, graphs were introduced as geometric representations of problem situations. To construct the graphs you plotted points. Some calculators and computers can construct graphs automatically. Any calculator or computer that constructs graphs will be referred to as a ***graphing utility***. Whenever we construct a graph, we see only the portion of the ***plane*** on which the graph is drawn. This portion of the plane is called a ***viewing window*** or ***viewing rectangle***. When you answered Investigation 6, you determined some of the possible limits on the viewing window. Notice that there is no one right answer for the viewing window. Your answers to Investigation 6 may differ from your group members. Whenever a graphing utility is used, the viewing window must be established. A common method is to use the following representations.

Xmin: the **left boundary** of the viewing window. Xmin should be smaller than any *actual* input value. This is the smallest input value in the window.

Xmax: the **right boundary** of the window. Xmax should be larger than any *actual* input value. This is the largest input value in the window.

Xscl: the number of units (distance) between tick marks on the input axis. Tick marks begin at the origin even if the origin is not in the viewing window. A rule of thumb is to use a Xscl that is approximately one–tenth of the difference between Xmax and Xmin.

Ymin: the **bottom boundary** of the window. Ymin should be smaller than any *actual* output value. This is the smallest output value in the window.

Ymax: the **top boundary** of the window. Ymax should be larger than any *actual* output value. This is the largest output value in the window.

Yscl: the number of units (distance) between tick marks on the output axis. Tick marks begin at the origin even if the origin is not in the viewing window. A rule of thumb is to use a Yscl that is approximately one–tenth of the difference between Ymax and Ymin.

The four values Xmin, Xmax, Ymin, Ymax make up the *boundaries* of the viewing window (Figure 1).

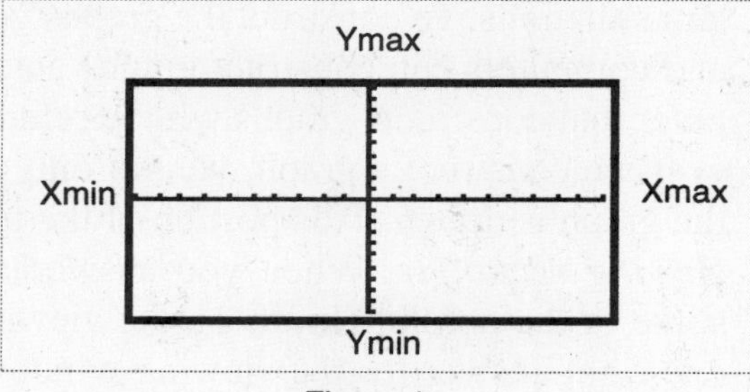

Figure 1

With reference to the previous investigation, the graph of Table 1 with the corresponding window settings appears in Figure 2. This is the graph of ***discrete*** points.

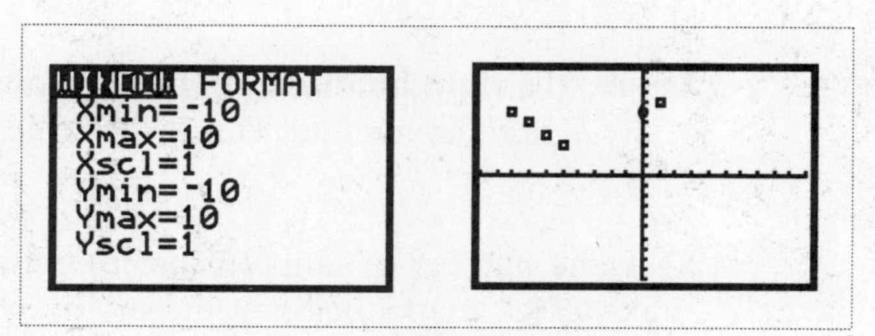

Figure 2

Many graphing utilities use a ***standard viewing window*** which is defined using the boundary values

Xmin = –10; Xmax = 10; Xscl = 1;
Ymin = –10; Ymax = 10; and Yscl = 1.

This is the viewing window used in Figure 2.

We see only the site of worship and one site of mathematical writings. Alter the viewing window so that all sites are visible (Figure 3).

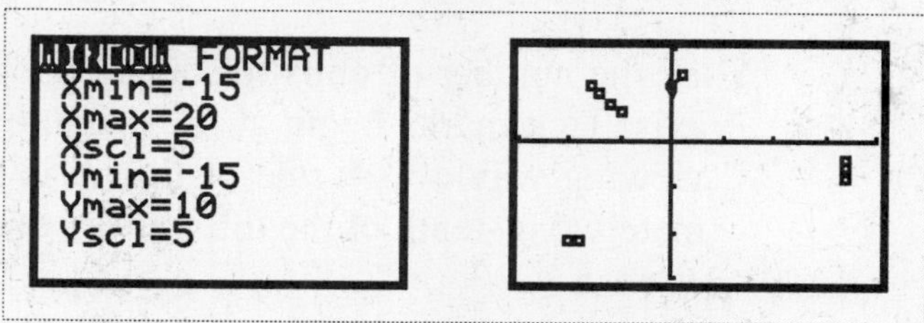

Figure 3

Investigation

8. The data in Table 2 displays the enrollment pattern at a particular college over a five year period.

Year	Number of students
1989	23969
1990	25186
1991	26866
1992	26588
1993	24196

Table 2: Enrollment data

Enter the data in your calculator, set an appropriate viewing window, and graph the data.

a. List the values used for the viewing window.

b. Sketch the graph below. Clearly label axes and points.

Do you remember Okimbe, our intrepid T–shirt salesperson? Table 3 displays the correspondence between T–shirts sold and profit.

Number of T–shirts sold	Profit ($)
0	–1000
50	–500
100	0
150	500
200	1000
250	1500

Table 3: T–shirts sold and profit

9. Refer to the data in Table 3. Identify

   a. the smallest input value

   b. the largest input value

   c. the smallest output value

   d. the largest output value.

10. a. Identify an appropriate viewing window to graph the data in Table 3. Record the window below.

    b. Enter the data into your calculator and construct the graph of discrete points.

Discussion

In Investigation 10, you created the graph from a table (the numeric representation of the relationship). The graph and a possible viewing window are displayed in Figure 4.

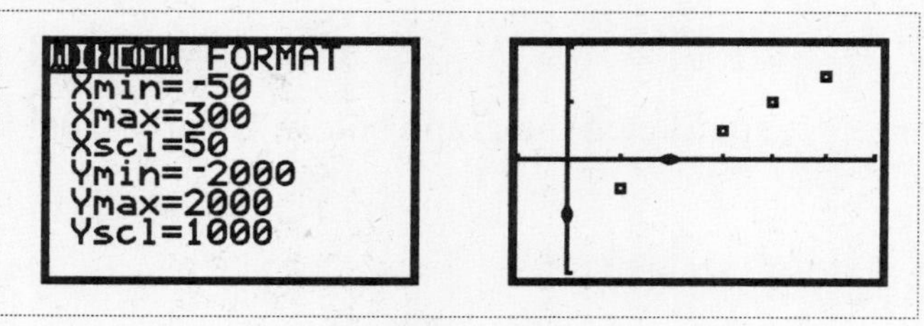

Figure 4

We can create the graph from the algebraic representation.

- Enter the algebraic representation of the function into your graphing utility. Ask your instructor or a classmate, or consult your manual to find out how to do this. Record the steps at the end of this section.

  and

- Plot the graph in the appropriate viewing rectangle. Ask your instructor or a classmate, or consult your manual to find out how to do this. Record the steps at the end of this section.

On many graphing utilities, y (or some similar variable) is used as a symbol for the output. When entering the algebraic representation, we only enter the rule involving the input variable. Many graphing utilities require that the input variable be x.

Investigation

11. For Table 3, the algebraic representation is $P(x) = 10x - 1000$ where x represents the number of T–shirts sold and P represents the profit. Graph this function in the same viewing rectangle used for Investigation 10.

12. How is the Investigation 11 graph different from the Investigation 10 graph?

13. Use **Trace** on your graphing utility to find and record three more ordered pairs on the function's graph.

14. If your calculator can display a table, compare and contrast the entries in the table with the values obtained by tracing along the graph. Which representation more accurately represents the problem situation?

15. Do the Investigation 13 answers make sense in the context of the input (the number of T–shirts sold) and in the context of the output (profit)?

Discussion

The ordered pairs from Investigation 13 imply that inputs which are not whole numbers were used. (Note the ordered pair displayed in Figure 5.) In

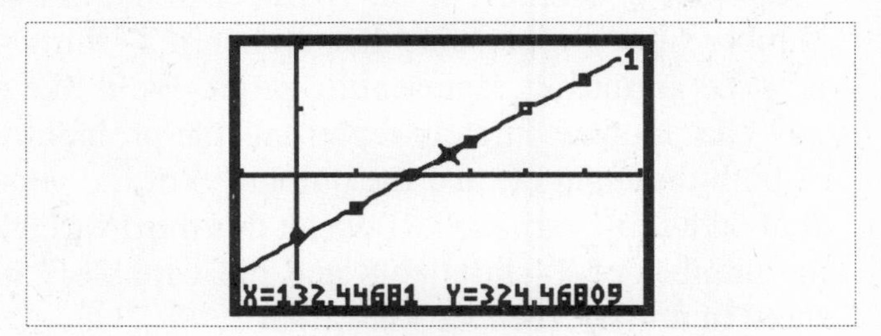

Figure 5

terms of the problem situation, the inputs should represent the number of T–shirts sold. The input must be a whole number (not 132.44681 as in Figure 5) to satisfy the problem situation. The graphing utility gave us a graph that contains all ordered pairs that satisfy the problem situation. Additionally, the graph contains points that satisfy the algebraic representation, but do not satisfy the problem situation. Whenever we construct a graph, we must be aware of this distinction when it occurs.

The connected graph created when the equation $P(x) = 10x - 1000$ was entered represents the relationship for an infinite number of table values. If we consider all points on the connected graph, we have ordered pairs that satisfy the algebraic representation, but do not satisfy the practical relationship. For example, the ordered pair (132.44681, 324.46809) is graphed as a point on the connected graph and satisfies the algebraic representation since $P(132.44681) = 324.46809$. But the pair does not satisfy the practical situation since it is not reasonable to sell 132.44681 T–shirts and make a profit of $324.46809. However, this point gives us a rough estimate of an actual pair of values that satisfy the relationship. We could say that Okimbe will make a profit of about $324 if he sells about 132 shirts. Why?

We call the connected graph a graph of the ***algebraic representation***. The points on the graph (whole number inputs, in this case) represent the graph of the ***problem situation***.

The set of inputs that produce outputs in a relationship is called the ***domain*** of the relationship. For the algebraic representation $P(x) = 10x - 1000$ or the corresponding connected graph, the domain is all numbers (we'll explain what this *really* means later). For the problem situation, the domain is all whole numbers.

The set of outputs in a relationship is called the ***range*** of the relationship. For the algebraic representation $P(x) = 10x - 1000$ or the connected graph, the range is all numbers. For the problem situation the range is the set of integers greater than or equal to –1000 that are multiples of 10.

The points you trace on the graph, such as (132.44681, 324.46809), do not always represent values in the problem situation. Okimbe cannot sell 132.44681 T–shirts. The first number in the ordered pair must be a whole number since it represents the number of T–shirts sold. The second number must be an integer representing profit. We'd like to create a viewing window that more accurately represents the problem situation. To do so, think of both the input (x) and the output (y) of the relationship as being dependent on a third variable t. If we let this third variable, as well as x, represent the number of T–shirts sold and the variable y represent the profit measured in dollars, then we know that

$$x = t$$

and

$$y = 10t - 1000.$$

When two variables x and y are defined in terms of a third variable t, the resulting equations are called ***parametric equations***. By doing this we can insure that the x–value is a whole number since we will control the value of t in this problem.

Investigation

16. Use the equations given in the previous discussion to complete Table 4

t	x	y
0		
50		
100		
150		
200		
250		

Table 4: Shirts sold and profit 2

17. Set your calculator to parametric mode. Ask your instructor or a classmate, or consult your manual to find out how to do this. Enter the equations.

18. To set the viewing window, you need to define Tmin, Tmax, and Tstep. To do so initially, use the smallest t value in Table 4 as Tmin, the largest t value in Table 4 as Tmax, and the change in t, (Δt), as Tstep. Use the same values for the other window settings that you used for Investigation 10. Graph the function and trace along the graph.

    a. How do the values of x and y compare to the values you found by tracing in Investigation 13?

    b. When tracing, which values make more sense in terms of the problem? Why?

19. Change Tstep to 10 and graph the relationship. Trace along the graph.

    a. How does tracing on this graph compare to tracing on the graph in Investigation 18?

    b. Interpret what is happening as you trace along the graph. Write your answer in terms of T–shirts sold.

20. Refer to the parametric representation of the T–shirt problem. Why did we use the variable t when both x and t represent the number of T–shirts sold by Okimbe?

Discussion

Using parametric equations, we can control the values that are used as input to the relationship. This allows us to avoid the decimal values we obtained when we traced along the graph given in Figure 5. This is particularly useful when the inputs to the relationship are whole numbers. We also can control the amount of change in the input variable as we move from one point to the next along the curve. When you answered Investigation 18, the Tstep was 50. When you trace, the input value (the number of T–shirts sold) increases by 50. In Investigation 19, the number of T–shirts sold increases by 10.

In conclusion, notice that every function between two quantities can be expressed parametrically. Let the independent variable be t. Define the input as t and the output as an expression in t that calculates the output.

For example, the concert ticket problem stated that each ticket sold was \$19. If x is the number of tickets sold and y is the total receipts, then

$$y = 19x.$$

The equations that define this relationship parametrically are

$$x = t$$

and

$$y = 19t$$

where t represents the number of tickets sold.

Figure 6a displays this relationship in its original form. Figure 6b displays the relationship parametrically.

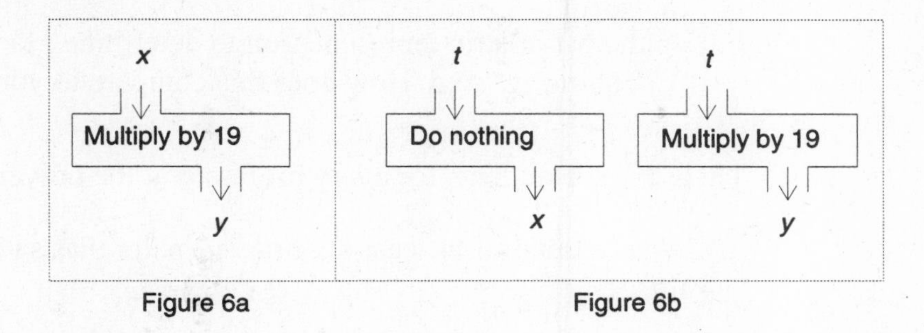

Figure 6a Figure 6b

Explorations

1. List and define words in this section that appear in ***italics bold*** type using your own words.

2. For Figure 7, identify the six values Xmin, Xmax, Xscl, Ymin, Ymax, and Yscl that constitute the viewing window.

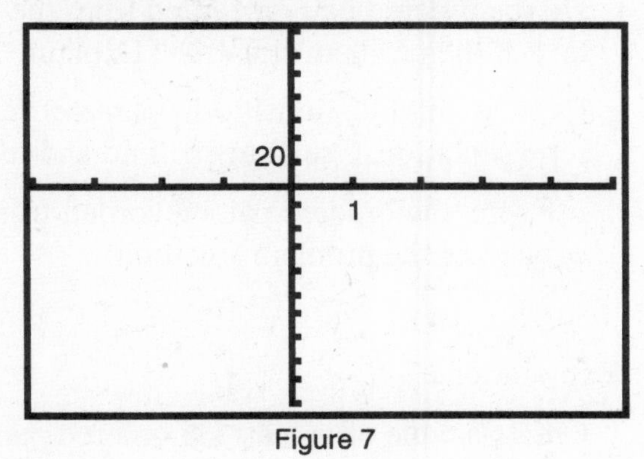

Figure 7

3. Set your graphing utility to the standard viewing window. In the graphing window, use the cross–hair cursor (+) to find

   a. the point closest to the ordered pair (7, 5). Record this ordered pair given on the graphing utility.

   b. the ordered pair nearest the ***origin*** (the intersection of the two axes). Record this ordered pair.

4. Try Xmin = –70 and Xmax = 400 for the graph of Okimbe's T–shirt profit function. Trace and compare the ordered pairs to those obtained in Figure 5. What do you notice?

5. Refer to the point identified in Figure 5.

   a. Substitute into the function to determine the actual profit if 132 shirts are sold? How does this compare to the output in Figure 5.

   b. Use the parametric representation to determine graphically the profit if 132 shirts are sold. How does this compare to your answer to a?

6. There are six students for every professor at the university.

   a. Create a table of at least six ordered pairs that satisfy this relationship.

   b. Enter the data from the table into your graphing utility.

   c. Set appropriate boundaries on the viewing window and construct the discrete plot.

   d. Create the algebraic representation (equation) for this problem. Enter the algebraic representation and graph in the same viewing window used for part c.

   e. Use **Trace** to find and record four more ordered pairs. Add these to your table.

   f. Do the ordered pairs recorded from the graph make sense in the context of the problem situation? Explain.

   g. Express this relationship in parametric form. Graph the parametric representation. Use **Trace** to find and record four more ordered pairs.

   h. Interpret the ordered pairs recorded from the parametric graph in the context of the problem situation.

For Explorations 7–9:

   a. Create a table of at least six ordered pairs.

   b. Construct a graph of discrete points in an appropriate viewing window.

   c. Graph the algebraic representation in the same viewing window used for part b.

   d. Use **Trace** to record three additional ordered pairs.

   e. Express the relationship in parametric mode.

   f. Graph in parametric mode and record three additional ordered pairs.

7. $C(z) = 2z - 7$

8. $A(b) = 3 + 2b^2$

9. $L(r) = 3r^2 - 2r + 7$

10. The graph in Figure 8 displays the position of an elevator as a function of time. Describe the motion of the elevator over each ten second period.

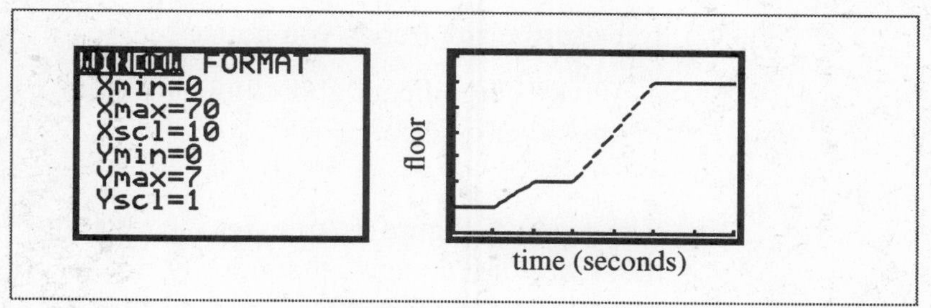

Figure 8

11. Set your calculator to DOT mode. Use parametric form to graph the factorial function on your graphing utility. Use boundaries of Tmin = 0, Tmax = 10, Tstep = 1, Xmin = 0, Xmax = 10, Xscl = 0, Ymin = –20, Ymax = 150, Yscl = 0.

    a. Describe the graph that you see.

    b. Use **Trace** to create a table of ten values for the factorial function.

    c. Change Tstep to 0.5. As you trace, what do you notice about the output values when the t is not a whole number?

12. The data in Table 5 displays the percentage of Americans with private health insurance as a function of the year.

Year	Percentage of Americans with private health insurance
1940	10
1950	55
1960	75
1970	80
1980	82
1990	65

Table 5: Private health insurance data
Source: Alliance for Health Reform

Enter the data in your calculator, set an appropriate viewing window, and graph the data.

a. List the values used for a viewing window.

b. Describe any trends you notice by looking at the graph.

c. What is significant about the data from 1980 to 1990 as opposed to any other decade?

13. The graph in Figure 9 expresses the relationship between the length and the width of rectangles that have perimeters of 20 feet.

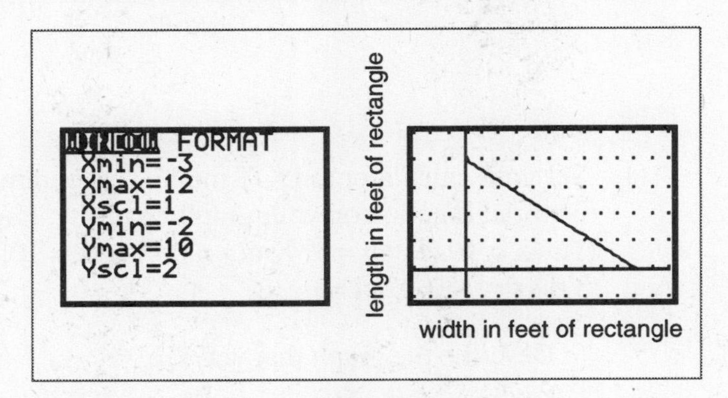

Figure 9

a. Record at least six ordered pairs.

b. Write a formula (equation) that expresses length as a function of width.

c. Does every point on the connected graph represent an ordered pair of values in the problem situation? Justify your answer.

Concept Map

Construct a concept map centered on the phrase **graphing utility**.

Reflection

Compare graphs constructed from a table (a numerical representation) and graphs constructed from an equation (an algebraic representation)? How are they different? How are they similar?

Describe the various features that you have learned to use on your graphing utility. How might you use each of these in the future?

Section 4.5 Functions over the Integers

Purpose

- Formalize the unary operations of oppositing and absolute value as functions.
- Compare and contrast the identity, the oppositing, and the absolute value functions.

Discussion

There have been three functions used in this chapter that merit further investigation. The first is the ***identity function***. This is the "do nothing" function that we used in Section 4.2. The identity function does nothing to the input. The output is the same as the input. The other two functions arose when we defined the integers. One is the oppositing function with output equal to the opposite of the input. The other is the absolute value function. If the input is non–negative, the absolute value function acts like the identity function. If the input is negative, the absolute value function acts like the oppositing function. Let's compare and contrast each of these functions numerically, algebraically, and graphically.

Investigation

1. Choose numbers and complete Table 1. Be sure to use both positive and negative inputs. Write an algebraic representation using x for the last row.

Input	Identity of input	Opposite of input	Absolute value of input
x			

Table 1: Numerical comparisons.

2. Recreate Table 1 using the table feature on your calculator if available.

3. Construct a function machine for the following. Clearly label the input, the process, and the output.

   a. The identity function.

   b. The oppositing function.

   c. The absolute value function.

4. a. Let $Ident(x)$ represent the function with input x and output the identity of x. Write an algebraic representation (equation) for $Ident(x)$.

   b. Let $Opp(x)$ represent the function with input x and output the opposite of x. Write an algebraic representation (equation) for $Opp(x)$.

   c. Let $Abs(x)$ represent the function with input x and output the absolute value of x. Write an algebraic representation (equation) for $Abs(x)$.

5. a. Graph the identity function. Describe the graph in words.

   b. Graph the oppositing function. Describe the graph in words.

   c. Graph the absolute value function. Describe the graph in words.

6. Graph both the identity function and the oppositing function in the same viewing window.

   a. Where are they the same? Where are they different?

   b. You are talking to a friend, Sarah, on the phone. Sarah knows what the graph of the identity function looks like. Explain to her how to obtain the graph of the oppositing function from the graph of the identity function.

   c. Use the **Trace** feature to find ordered pairs that are the same on both graphs, that have the same x but different y, and that have the same y but different x.

   d. Use the table feature of your calculator to find ordered pairs that are the same on both graphs, that have the same x but different y, that have the same y but different x.

7. Graph both the identity function and the absolute value function in the same viewing window.

   a. Where are they the same? Where are they different?

   b. You are talking to a friend, Ted, on the phone who knows what the graph of the identity function looks like. Explain how to obtain the graph of the absolute value function from the graph of the identity function.

c. Use the **Trace** feature to find ordered pairs that are the same on both graphs, that have the same x but different y, and that have the same y but different x.

d. Use the table feature of your calculator to find ordered pairs that are the same on both graphs, that have the same x but different y, that have the same y but different x.

8. Graph both the oppositing function and the absolute value function in the same viewing window.

a. Where are they the same? Where are they different?

b. You are talking to a friend, Terra, on the phone who knows what the graph of the oppositing function looks like. Explain how to obtain the graph of the absolute value function from the graph of the oppositing function.

c. Use the **Trace** feature to find ordered pairs that are the same on both graphs, that have the same x but different y, and that have the same y but different x.

d. Use the table feature of your calculator to find ordered pairs that are the same on both graphs, that have the same x but different y, that have the same y but different x.

Discussion

We have looked at the functions $Ident(x)$, $Opp(x)$, and $Abs(x)$ in the preceding investigations.

Table 1 illustrates the numeric representation of these functions. The function machines for each appear in Figure 1.

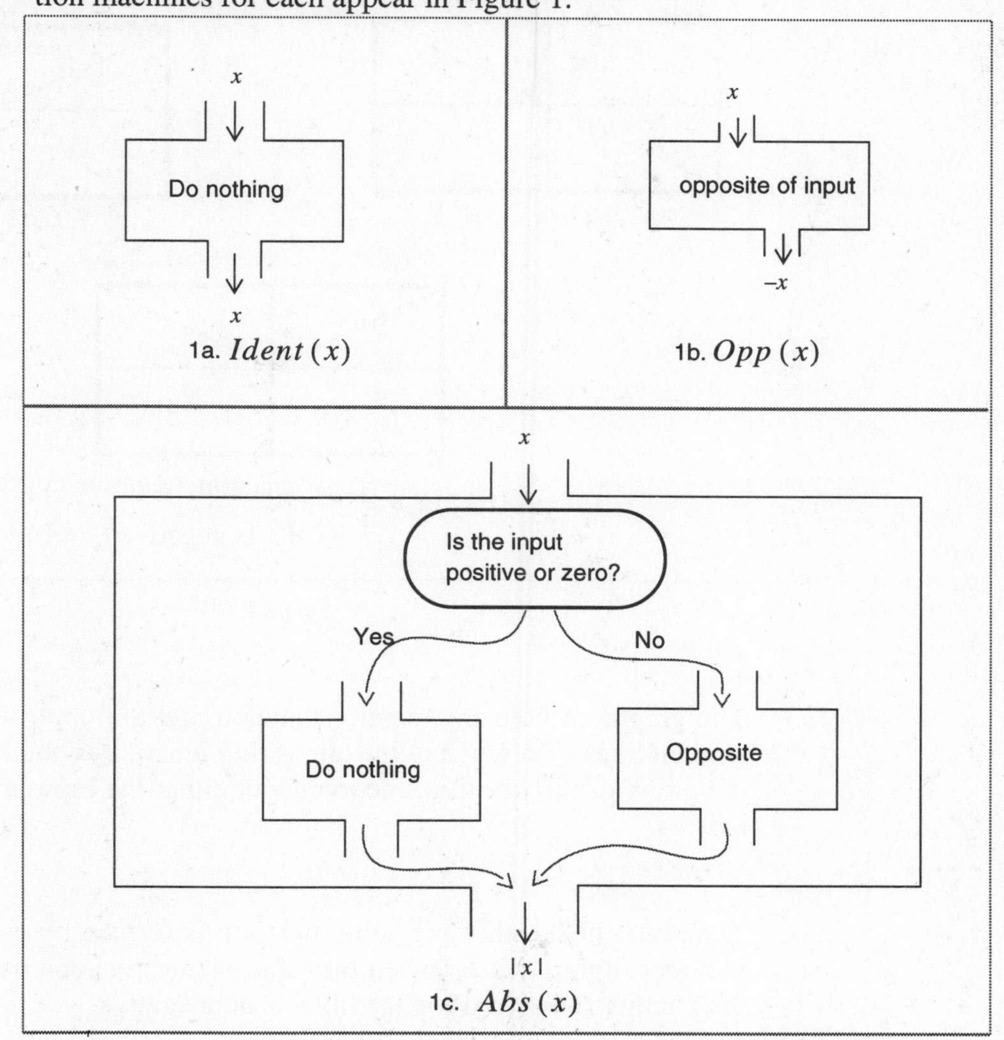

Figure 1

The algebraic representations are

- $Ident\,(x) = x$
- $Opp\,(x) = -x$
- $Abs\,(x) = \begin{cases} x \text{ if } x \text{ is non–negative } (x \geq 0) \\ -x \text{ if } x \text{ is negative } (x < 0) \end{cases}$.

The geometric representations appear in Figure 2 a, b, and c.

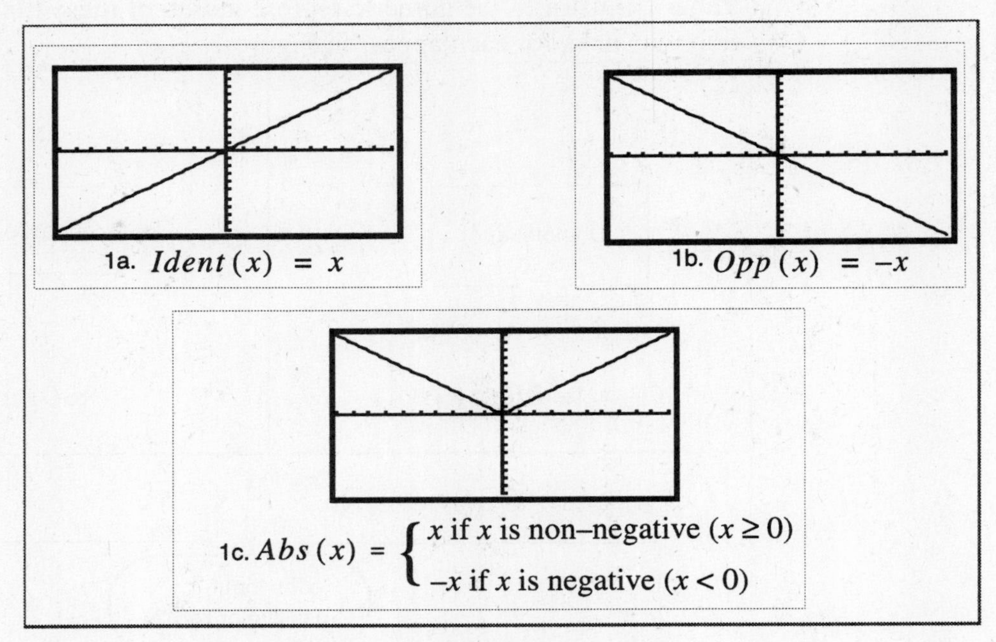

Figure 2

The graphs of both the identity function and the oppositing function are straight lines. The graph of the oppositing function is obtained by reflecting the graph of the identity function about either the input axis or the output axis.

The graph of the absolute value function is formed by using the graph of the oppositing function when the input is negative and using the graph of the identity function when the input is nonnegative.

Let's apply these ideas to more complex functions.

Investigation

Let function M be defined by $M(x) = 3x - 2$.

Then define the function *OppM* to be the opposite of function M and define the function *AbsM* to be the absolute value of function M.

9. Choose numbers and complete Table 2. Be sure to use both positive and negative inputs. Write an algebraic representation using x for the last row.

Input	$M(x)$	$OppM(x)$	$AbsM(x)$
x			

Table 2: Numerical comparisons.

10. Recreate Table 2 using the table feature on your calculator if available.

11. Construct function machines for the functions $M(x)$, $OppM(x)$, and $AbsM(x)$.

12. Write the algebraic representation of the functions:

a. $M(x) =$

b. $OppM(x) =$

c. $AbsM(x) =$

13. a. Graph the functions $M(x)$, $OppM(x)$, and $AbsM(x)$ in the standard viewing rectangle.

b. Explain how to use the graph of $M(x)$ to obtain the graph of $OppM(x)$.

c. Explain how to use the graphs of $M(x)$ and $OppM(x)$ to obtain the graph of $AbsM(x)$.

Discussion

Table 3 is a numeric representation of functions $M(x)$, $OppM(x)$, and $AbsM(x)$.

Input	$M(x)$	$OppM(x)$	$AbsM(x)$
–3	–11	11	11
–2	–8	8	8
–1	–5	5	5
0	–2	2	2
1	1	–1	1
2	4	–4	4
3	7	–7	7

Table 3: Numerical comparisons.

If $M(x) = 3x - 2$ then

$$OppM(x) = -(3x-2) \text{ and } AbsM(x) = |3x-2|.$$

The graphs of each appear in Figure 3.

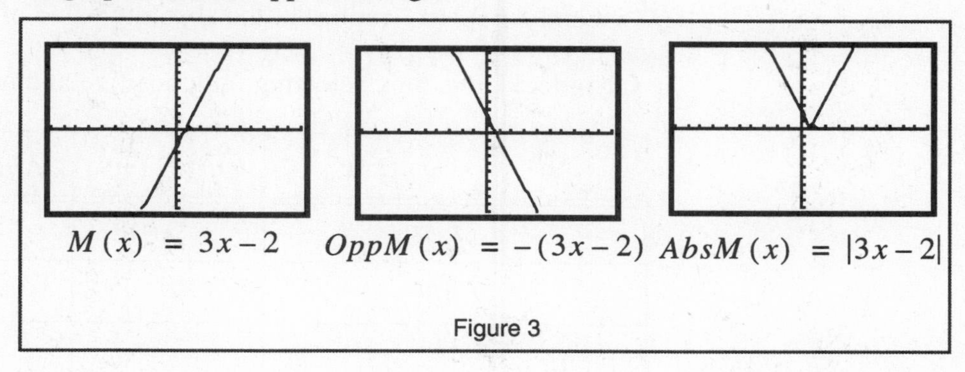

Figure 3

Explorations

1. List and define words in this section that appear in ***italics bold*** type using your own words.

2. Given $L(x) = 2x$. Let $OppL(x)$ represent the opposite of $L(x)$ and $AbsL(x)$ represent the absolute value of $L(x)$.

   a. Complete Table 4 by selecting both positive and negative inputs.

x	$L(x)$	$OppL(x)$	$AbsL(x)$

Table 4: *L*(*x*), *OppL*(*x*), and *AbsL*(*x*)

   b. Write algebraic representations for $OppL(x)$ and $AbsL(x)$.

   c. Graph all three functions.

   d. Describe how the graphs of $L(x)$ and $OppL(x)$ are related.

   e. Describe how the graphs of $L(x)$ and $AbsL(x)$ are related.

3. Given $Q(x) = x^2 - 5x - 3$. Let $OppQ(x)$ represent the opposite of $Q(x)$ and $AbsQ(x)$ represent the absolute value of $Q(x)$.

a. Complete Table 5 by selecting both positive and negative inputs.

x	$Q(x)$	$OppQ(x)$	$AbsQ(x)$

Table 5: $Q(x)$, $OppQ(x)$, and $AbsQ(x)$

b. Write algebraic representations for $OppQ(x)$ and $AbsQ(x)$.

c. Graph all three functions.

d. Describe how the graphs of $Q(x)$ and $OppQ(x)$ are related.

e. Describe how the graphs of $Q(x)$ and $AbsQ(x)$ are related.

4. Let $G(x) = -(3x - 5)$ and $H(x) = -3x + 5$.

a. Complete Table 6 by selecting both positive and negative inputs.

x	$G(x)$	$H(x)$

Table 6: $G(x)$ versus $H(x)$

b. Construct function machines for both functions.

c. Graph both functions in the standard viewing rectangle.

d. Describe the relationship between the two graphs.

5. Let $G(x) = -(x^2 - 5x - 3)$ and $H(x) = -x^2 + 5x + 3$.

   a. Complete Table 7 by selecting both positive and negative inputs.

x	$G(x)$	$H(x)$

Table 7: $G(x)$ **versus** $H(x)$

   b. Graph both functions in the standard viewing rectangle.

   c. Describe the relationship between the two graphs.

6. Let $G(x) = -(3x^2 + 5x - 1)$ and $H(x) = -3x^2 - 5x + 1$.

   a. Complete Table 8 by selecting both positive and negative inputs.

x	$G(x)$	$H(x)$

Table 8: $G(x)$ **versus** $H(x)$

   b. Graph both functions in the standard viewing rectangle.

   c. Describe the relationship between the two graphs.

7. Assume x represents a negative number.

   a. What is the sign of x?

   b. What is the sign of $-x$?

   c. What is the sign of $|x|$?

   d. Select a number for x. Then write numeric values of $-x$ and $|x|$?

8. Based on the results of Explorations 4– 6, write a function with no parentheses that is equivalent to the functions in a and b below. Check your answer numerically and graphically.

   a. $f(x) = -(9 + 2x)$.

   b. $f(x) = -(3x^2 - 4x + 7)$.

9. For each given function $f(x)$ in Exploration 8, will the output ever be negative? Justify your answer by providing a numerical example.

10. Based on Explorations 4–6, use parentheses and oppositing to write a function that is equivalent to the functions in a and b below.

    a. $fun(x) = 13x - 4$

    b. $joy(x) = 3 - 2x - x^2$

11. The Illini scored 30 points in the first half of a basketball game. They did it without scoring a single foul shot. How did they score the 30 points using just 2 and 3 point shots. Be sure to justify your answer(s).

12. Are the expressions $3 + 2m$ and $5m$ equivalent? Explain.

13. a. Write an expression for the perimeter of a rectangle that has a width of 4 units and a length of k units.

    b. Write an expression for the area of the rectangle described in part a.

Concept Map

Construct a concept map centered on the phrase **absolute value function**.

Reflection

How do you enter an opposite on your calculator? How do you enter an absolute value on your calculator?

Section 4.6 Making Connections: How do Integers Expand the Mathematics ?

Purpose

- Reflect upon ideas explored in Chapter 4.
- Explore the connections among representations of functions.

Investigation

In this section you will work outside the system to reflect upon the mathematics in Chapter 4: what you've done and how you've done it.

1. State the five most important ideas in this chapter. Why did you select each?

2. Identify all the mathematical concepts, processes and skills you used to investigate the problems in Chapter 4?

3. What has been common to all of the investigations which you have completed?

4. Select a key idea from this chapter. Write a paragraph explaining it to a confused best friend.

5. You have investigated many problems in this chapter.

a. List your three favorite problems and tell why you selected each of them.

b. Which problem did you think was the most difficult and why?

6. Explain what happens to x as it is input into an absolute value function machine.

Discussion

There are a number of really important ideas which you might have listed including integer, operations on the integers, properties of integers, the absolute value, domain, range, piecewise functions and additive inverses.

Concept Map

Construct a concept map centered around the phrase **discrete functions.**

Reflection

How does the absolute value function differ from other functions we have studied and how does domain play a different role?

Illustration

Draw a picture of **a mathematics practioner** (a person who uses mathematics consciously as part of his/her job).

A MATHEMATICS
PRACTIONER

Chapter 5

Rational Numbers: Further Expansion of a Mathematical System

Section 5.1 Rational Numbers and the Algebraic Extension

Purpose

- Introduce the rational number system.
- Investigate operations with rational numbers.
- Explore multiple representations of rational numbers.
- Introduce the reciprocals of the integers.

Investigation

Table 1 displays the departure time, the arrival time, and the frequent flyer miles for selected trips as reported by the airlines.

Trip (From–To)	Departure time	Arrival time	Frequent flyer miles
Chicago–Boston	8:00 a.m.	11:14 a.m.	886
Los Angeles–Chicago	1:20 p.m.	7:23 p.m.	1853
Seattle–Minneapolis	11:50 a.m.	4:54 p.m.	1462
Phoenix–Dallas	3:50 p.m.	7:15 p.m.	927
Miami–New York	5:40 p.m.	8:30 p.m.	1124
Atlanta–Little Rock	11:56 a.m.	12:20 p.m.	462
Detroit–San Francisco	9:15 a.m.	11:27 a.m.	2301

Table 1: Airline travel data

1. Complete Table 2 by recording the elapsed time (beware of time zone changes) for each flight. Express the time in hour–minute format, decimal form, and fractional form.

Trip (From–To)	Time (hour–minute)	Time (decimal)	Time (fraction)
Chicago–Boston			
Los Angeles–Chicago			
Seattle–Minneapolis			
Phoenix–Dallas			
Miami–New York			
Atlanta–Little Rock			
Detroit–San Francisco			

Table 2: Airline travel times

2. Assuming that the frequent flyer miles represent the approximate distance traveled, determine how far each plane travels in one hour.

Trip (From–To)	Frequent flyer miles	Elapsed time	Distance traveled in one hour
Chicago–Boston			
Los Angeles–Chicago			
Seattle–Minneapolis			
Phoenix–Dallas			
Miami–New York			
Atlanta–Little Rock			
Detroit–San Francisco			

Table 3: Airline travel distances per hour

3. What process did you use to answer the previous investigation? Which representation for elapsed time seems best for this computation?

4. Consider sharing a pizza. The amount you get is related to the number of "friends" who get to share it.

   a. If P represents the amount of pizza you have, tell how much pizza each person gets (remember the concept of fair share) if the number of people to share the pizza is 1, 3, 4, or 12.

   b. Draw a picture of the amount of pizza P represents.

   c. Begin each problem with the same picture and show one person's share of pizza if there are 1, 3, 4, or 12 people. What if there are 0 people?

5. What type of number is needed to answer Investigation 1? How is this type of number different from any preceding number we have studied?

6. The whole numbers were not closed under two operations. What were the two operations? Give an example for each showing why.

7. Which of the two operations now have solutions if we can use fractions?

Discussion

Each of the elapsed times for the trips listed in Table 1 involve fractional parts of an hour. These are represented three ways: as a number of minutes; as a fraction by using the number of minutes as the ***numerator*** and sixty as the ***denominator***; and, as the decimal corresponding to dividing the number of minutes by sixty.

For example, the Chicago to Boston trip has an elapsed time of two hours and fourteen minutes, written 2:14. Expressed as a fraction, this number is $2\frac{14}{60}$ or $2\frac{7}{30}$ in reduced form. The decimal representation is 2.2333333....

Thus the set of integers is no longer sufficient to represent this set of elapsed times. The integers are closed under addition, subtraction, and multiplication, but not under division. This is true since the integers do not include fractions or decimals. The integers are primarily used to answer questions that ask "how many". Once we have fractions or decimals, we can answer questions about measurement, such as time, rather than counts.

To complete Table 3, we must divide the total distance of the trip by the elapsed time. The time must be in either fractional or decimal form to perform this division. Using the decimal form, we can perform the division on the calculator. For example, the distance travelled in one hour on the Chicago–Boston flight is computed as $886 \div 2.2333333...$. This is approximately equal to 397 miles in one hour. When performing approximate computations, we use the symbol $\approx$ to represent "approximately equal". So we would write $886 \div 2.2333333... \approx 397$.

To perform the division with the fractional representation, we need to explore the idea of ***reciprocal***. We'll look at this a little later in this section.

Let's look at the investigations you did involving the pizza.
The picture you drew for the amount of pizza could be one of two types. One huge pizza and the representation of one person's share when there were 4 people looks like Figure 1.

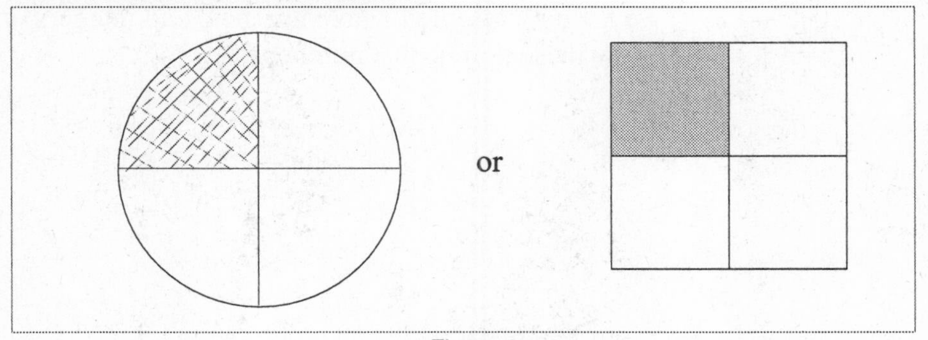

Figure 1

In both examples you have divided a unit into the appropriate number of pieces and chosen one of those pieces. The reciprocal of a number is used to represent the idea of dividing a whole into equal parts and taking one part. So the reciprocal of 4, $\frac{1}{4}$, is related to the picture as dividing into how many parts (4) and taking one part.

If you really like pizza you may have chosen to represent the amount of pizza available as several pizzas—especially knowing you have to share it among 12 people. Suppose you draw twelve individual pizzas then the problem of sharing it among 4 people looks like Figure 2.

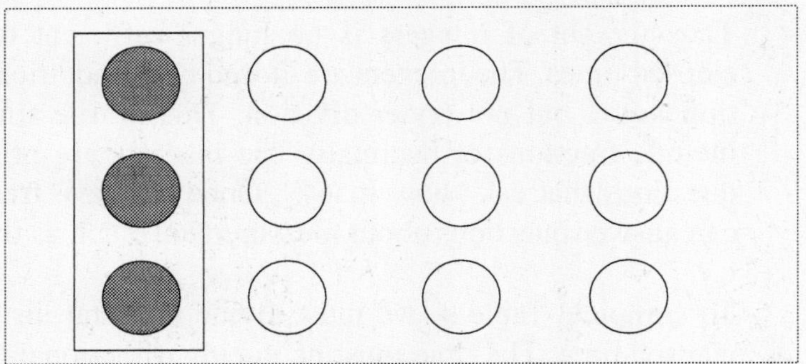

Figure 2

Because our unit is a set of pizzas, $\frac{1}{4}$ of P looks like three pizzas or $\frac{3}{12}$ of the total set of pizza. Thus we have encountered an unusual property of fractions—there are many representations of $\frac{1}{4}$ which are equivalent. All of the fractions $\frac{2}{8}, \frac{3}{12}, \frac{5}{20}, \frac{25}{100}, \frac{40}{160}$ are equivalent to $\frac{1}{4}$.

Now that we have seen some uses of fractions and decimals, let's investigate them in more depth.

Investigation

8. We have represented whole numbers and integers on a number line. Show these points on a number line.

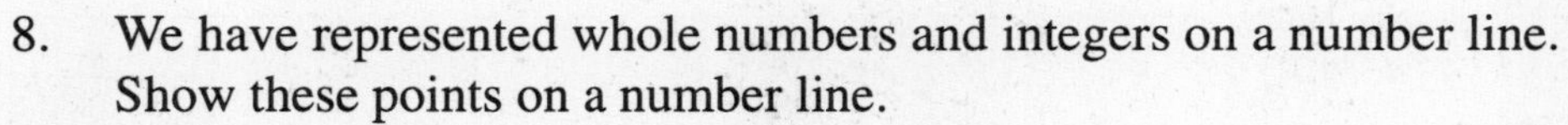

a. 0 b. 1

c. –1 d. $\frac{1}{2}$

e. $\frac{3}{4}$ f. $\frac{1}{3}$

g. $1\frac{1}{2}$ h. $\frac{5}{3}$

i. $-\frac{1}{4}$

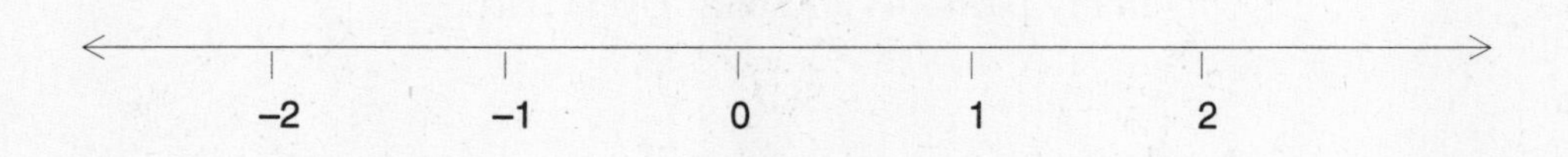

9. Describe how you located the point $\frac{3}{4}$ on the number line.

10. Consider $\frac{3}{4}$.

    a. How could $\frac{3}{4}$ be considered a process?

    b. Explain why we say that the symbol $\frac{3}{4}$ represents one number on the number line. Explain how this is an output.

    c. Give another reason why $\frac{3}{4}$ is a single number.

Discussion

If we divide integers we can represent this process as $4 \div 2$ or $2\overline{)4}$ or $\frac{4}{2}$.

So $\frac{1}{4}$ is the answer to $1 \div 4$ because both represent the same process.

We need to expand our number system to include fractions and the corresponding decimals. The collection of all numbers of the form **integer divided by non–zero integer** is called the set of ***rational numbers***. If we use **Q** to represent the set of rational numbers then

$$\mathbf{Q} = \{\tfrac{a}{b} \text{ where } a \text{ and } b \text{ are integers and } b \text{ is not zero}\}.$$

Notice that the integers are a subset of the rational numbers since the integers are the set of rational numbers in which the denominator is 1.

In order to use fractions to calculate the distance travelled in one hour for each of the trips listed in Table 1, we must understand the idea of reciprocal. The next investigation will provide some insight.

Investigation

11. Complete Table 4 using the inverse key x^{-1} on your calculator to find the reciprocal of each number.

Number	Reciprocal	Product of number and reciprocal
$\frac{2}{5}$		
$\frac{1}{4}$		
$-\frac{2}{3}$		
3		
−13		
$3\frac{1}{3}$		
0.75		
2.5		
6.84		

Table 4: Finding reciprocals with a calculator

12. Describe the result if you multiply a number and its reciprocal.

13. Is there any number that does not have a reciprocal? If so, identify it and explain why.

14. Complete Table 5 without using a calculator. Express each reciprocal in fractional form.

Number	Reciprocal	Product of number and reciprocal
$\frac{2}{5}$		
$\frac{1}{4}$		
$-\frac{2}{3}$		
3		
−13		
$3\frac{1}{3}$		

Table 5: Finding reciprocals without a calculator

15. Given a fraction, briefly describe how to find the reciprocal of the fraction **as a fraction**.

16. Decimal representations of fractions can be done on your calculator. Complete Table 6.

Fraction	Decimal (Guess)	Decimal (Calculator)
$\frac{1}{2}$		
$\frac{1}{3}$		
$\frac{1}{4}$		
$\frac{2}{3}$		

Table 6: Decimal Representation of Fractions

Fraction	Decimal (Guess)	Decimal (Calculator)
$\frac{1}{5}$		
$1\frac{1}{2}$		
	0.75	
		1.3333333
$\frac{0}{3}$		

Table 6: Decimal Representation of Fractions

Discussion

The reciprocal of a rational number is called the ***multiplicative inverse*** of the rational number. One characteristic about a rational number and its multiplicative inverse is that their product is one, the ***multiplicative identity***.

If a represents a rational number, the multiplicative inverse, or reciprocal, of a is represented by a^{-1}.

For example, if $a = 17$ then $a^{-1} = (17)^{-1} = \frac{1}{17}$.

Also, if $a = \frac{1}{3}$ then $a^{-1} = \left(\frac{1}{3}\right)^{-1} = 3$.

When finding a reciprocal, we operate on one number to obtain a second number. The operation of finding a multiplicative inverse, symbolized by a raised –1, is an ***unary operation*** (unary since it operates on only one number at a time).

You can see from Table 6 that the calculator gives us a decimal representation of any fraction in which the numerator and denominator are integers and the denominator is not zero.

These decimals fall into two types, terminating and repeating, which may be hard to determine from a calculator.

We now use the reciprocal operator to investigate the relationship between division and multiplication. This will allow us to calculate the distance travelled in one hour for the trips in Table 1.

Investigation

17. Compute. Compare the results of each division with the multiplication adjacent to it.

a. $8 \div 4 =$ ______ $8\left(\frac{1}{4}\right) =$ ______

b. $7 \div 3 =$ ______ $7\left(\frac{1}{3}\right) =$ ______

c. $-5 \div \frac{1}{7} =$ ______ $-5\,(7) =$ ______

d. $3 \div \left(-\frac{6}{7}\right) =$ ______ $3\left(-\frac{7}{6}\right) =$ ______

18. a. Rewrite $5 \div 2$ as a multiplication.

b. Rewrite $a \div b$ as a multiplication.

c. Describe how to rewrite a division as a multiplication.

19. Table 7 contains the elapsed times, in fractional form, for each of the airline trips. Complete the table by writing the reciprocal, in fractional form, of each elapsed time.

Trip (From–To)	Elapsed time	Reciprocal of elapsed time
Chicago–Boston	$2\frac{7}{30}$	
Los Angeles–Chicago	$4\frac{1}{20}$	
Seattle–Minneapolis	$3\frac{1}{15}$	
Phoenix–Dallas	$2\frac{5}{12}$	
Miami–New York	$2\frac{5}{6}$	
Atlanta–Little Rock	$1\frac{2}{5}$	
Detroit–San Francisco	$5\frac{1}{5}$	

Table 7: Airline travel times

20. Record the reciprocals of the elapsed times from Table 7 in Table 8. Use the distance and the reciprocal of the elapsed time to calculate the distance traveled in one hour.

Trip (From–To)	Frequent flyer miles	Reciprocal of elapsed time	Distance traveled in one hour
Chicago–Boston			
Los Angeles–Chicago			
Seattle–Minneapolis			
Phoenix–Dallas			
Miami–New York			
Atlanta–Little Rock			
Detroit–San Francisco			

Table 8: Airline travel distances per hour

21. Explain how the reciprocal of the elapsed time can be used to calculate the distance traveled in one hour.

22. Explain why division by a fraction can be accomplished by using multiplication of the reciprocal of the divisor.

Discussion

We see that division ***is equivalent to*** multiplying by the reciprocal of the ***divisor***. This is accomplished by making two changes.

- Change the operation from division to multiplication.
- Change the divisor to its reciprocal (Figure 3).

Change operation

$$a \div b = a \cdot b^{-1}$$

Change b to its reciprocal b^{-1}

Figure 3

Explorations

1. List and define words in this section that appear in ***italics bold*** type using your own words.

2. Find the reciprocal.

   a. 6   b. –5

   c. $-\frac{2}{3}$ d. 3.92

   e. $7\frac{9}{11}$ f. $5x$

   g. $-3y$ h. $x + y$

3. Write each of the following as a multiplication.

   a. $5 \div 17$ b. $-3 \div \frac{1}{11}$

   c. $-2.3 \div \frac{5}{6}$ d. $\frac{4}{7} \div (-2)$

   e. $-9.2 \div -2\frac{1}{5}$ f. $x \div 3y$

   g. $5t \div \frac{1}{q}$ h. $\frac{3}{k} \div \frac{d}{c}$

4. Describe in words how to add two rational numbers in

   a. fractional form.

   b. decimal form.

5. Describe in words how to subtract two rational numbers in

   a. fractional form.

   b. decimal form.

6. Describe in words how to multiply two rational numbers in

   a. fractional form.

   b. decimal form.

7. Evaluate each of the following.

a. $\left(\frac{7}{8}\right)\left(\frac{2}{3}\right)$

b. $\frac{7}{5} \div \left(-\frac{42}{17}\right)$

c. $-\frac{5}{8} + \frac{1}{12}$

d. $\frac{9}{10} - \left(-\frac{3}{4}\right)$

e. $\left(-\frac{2}{9}\right)\left(\frac{11}{5}\right)$

f. $-7\frac{3}{4} - 4\frac{2}{3}$

g. $\frac{5}{8} + \left(-\frac{7}{10}\right)$

h. $\left(-\frac{13}{12}\right) \div \left(-\frac{5}{9}\right)$

i. $\frac{2}{9} + \left(-\frac{7}{12}\right)$

j. $\frac{7}{9} - \left(\frac{5}{3}\right)\left(-\frac{3}{7}\right)$

8. Can you graph the rational numbers on a number line? Defend your answer.

9. a. Find a rational number between $\frac{1}{4}$ and $\frac{1}{3}$. (Hint: Find the average of the two numbers.)

   b. Find a rational number between $\frac{1}{8}$ and $\frac{1}{7}$.

   c. For any two rational numbers, is there a rational number between them? Justify.

   d. For any two integers, is there an integer between them? Justify.

10. a. Complete Table 9. Record the decimal form of the reciprocal.

Number	Reciprocal
8	
7	
6	
5	
4	
3	
2	

Table 9: More reciprocal exploration

   b. From Table 9, what value does the reciprocal seem to approach as the number gets closer and closer to 1?

11. Given $T(r) = 5r - 2$. Find $T(\frac{3}{5})$.

12. Given $P(x) = 3x^2 - x - 4$. Find $P(-\frac{2}{7})$.

13. a. The integers are closed under what operations?

    b. The rational numbers are closed under what operations?

14. For every two students who are full–time there are seven students who are part–time.

    a. Express the relationship between the number of full–time students and the number of part–time students as a proportion.

    b. If $F(P)$ represents the number of full–time students F as a function of the number of part–time students P, write the algebraic representation.

    c. Create a table of at least five input/output pairs.

    d. Draw a graph of the relationship.

15. A store has marked down all its merchandise 27% to celebrate its 27th anniversary.

    a. Let P be the original price and S be the sale price. Create a table that uses the original price as input and the sale price as output. Fill in the table with at least 5 input/output pairs.

    b. Write an algebraic representation for $S(P)$.

    c. Draw a graph of the relationship.

Concept Map

Construct a concept map centered on the phrase **rational numbers.**

Reflection

Write an example of a real–life function in which fractions or decimals are used. What is the input? What is the output?

Section 5.2 Rates of Change

Purpose

- Explore constant rates of change.
- Explore variable rates of change.
- Introduce slope.

Investigation

One morning exactly at sunrise, a monk began to climb a mountain. The graph in Figure 1 displays the relationship between the horizontal distance he travelled and the corresponding change in altitude. There is a third distance to consider–the length of the path. We'll consider this in a later section.

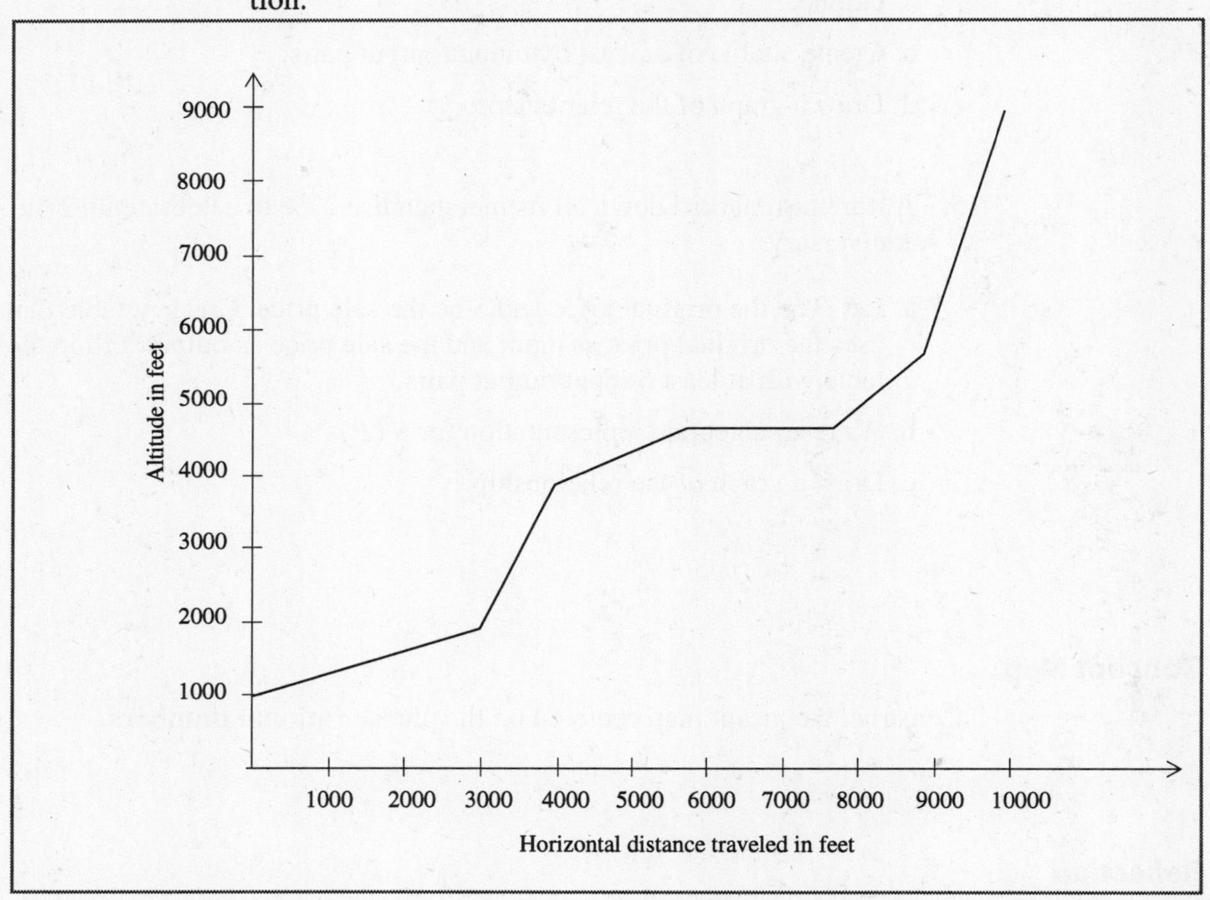

Figure 1

1. What points would you choose as interesting points to describe this hill? Label the points using A, B, C, . . . and write each as an ordered pair. Defend why you chose these points.

2. What would be the easiest part to walk? What would be the hardest part to walk? Why?

3. How would you find a mathematical expression or number to compare parts of the climb.

4. Describe why walking the first 3000 feet in distance might be easier than walking the distance from 3000 feet to 4000 feet.

5. a. Describe why going a distance of 1000 feet horizontally might be *easier* than walking 1000 feet vertically.

   b. Describe why going a distance of 1000 feet horizontally might be *harder* than walking 1000 feet vertically.

Discussion

In order to make comparisons using finite differences, you subtract values of elevation and values of horizontal distance. However the ratio of these differences gives more relevant information. For example, if walking up a hill is harder than walking the same distance on a level surface, the change in both horizontal distance and elevation must be considered in your computation.We can compare these by looking at the mountain's slopes—it is even called ***slope*** in mathematics.

Investigation

Consider Figure 2 in which the graph of Figure 1 has now been labelled.

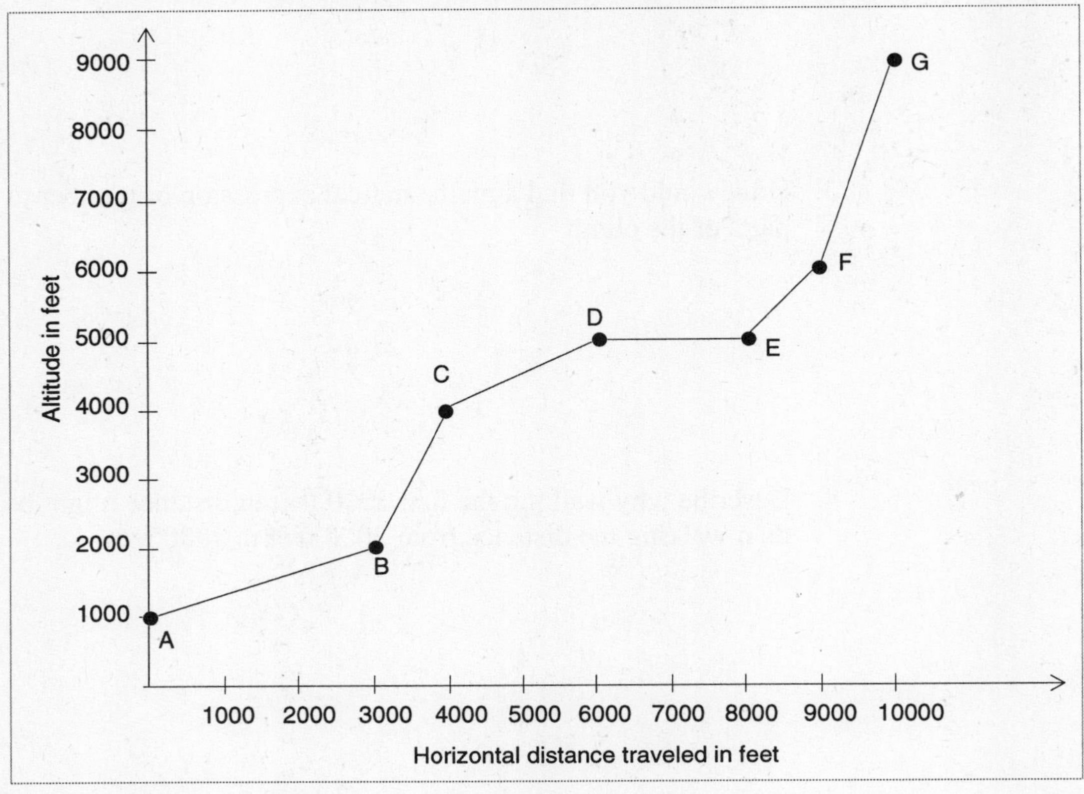

Figure 2

7. Complete Figure 3 recording the values from Figure 2 for horizontal distance traveled (the first coordinate of the ordered pairs) and altitude (the second coordinate of the ordered pairs). Find the finite differences for the horizontal distance traveled column and the altitude column.

Pt	Horizontal distance d traveled (ft.)	Finite Difference Δd	Altitude a (ft.)	Finite Difference Δa
A				
B				
C				
D				
E				
F				
G				

Figure 3

(Recall that Δ means "change in". See Section 4.2.)

8. Complete Table 1 by finding the ratio $\frac{\Delta a}{\Delta d}$ for each finite difference in Figure 3.

Endpoints	Δd	Δa	$\frac{\Delta a}{\Delta d}$
A to B			
B to C			
C to D			
D to E			
E to F			
F to G			

Table 1: Ratios of change in altitude to change in distance

9. Interpret in words the meaning of $\frac{\Delta a}{\Delta d}$ with respect to the monk's journey.

Discussion

We have investigated the change in each dimension as we moved from one point to another along the monk's trip. The change in horizontal distance traveled, Δd, varies from 1000 to 3000 feet as we move between points. The change in altitude, Δa, varies between 0 and 3000 feet as we move from point to point. We need to consider both changes to measure the amount of effort exerted as we move from point to point.

The ratio of changes is called the ***rate of change***.

In this case, $\frac{\Delta a}{\Delta d}$ represents the rate of change of altitude with respect to change in horizontal distance traveled. The larger this value, the steeper the path. The smaller the value, the more gradual the climb. Figure 4 displays the rates of change for each portion of the trip.

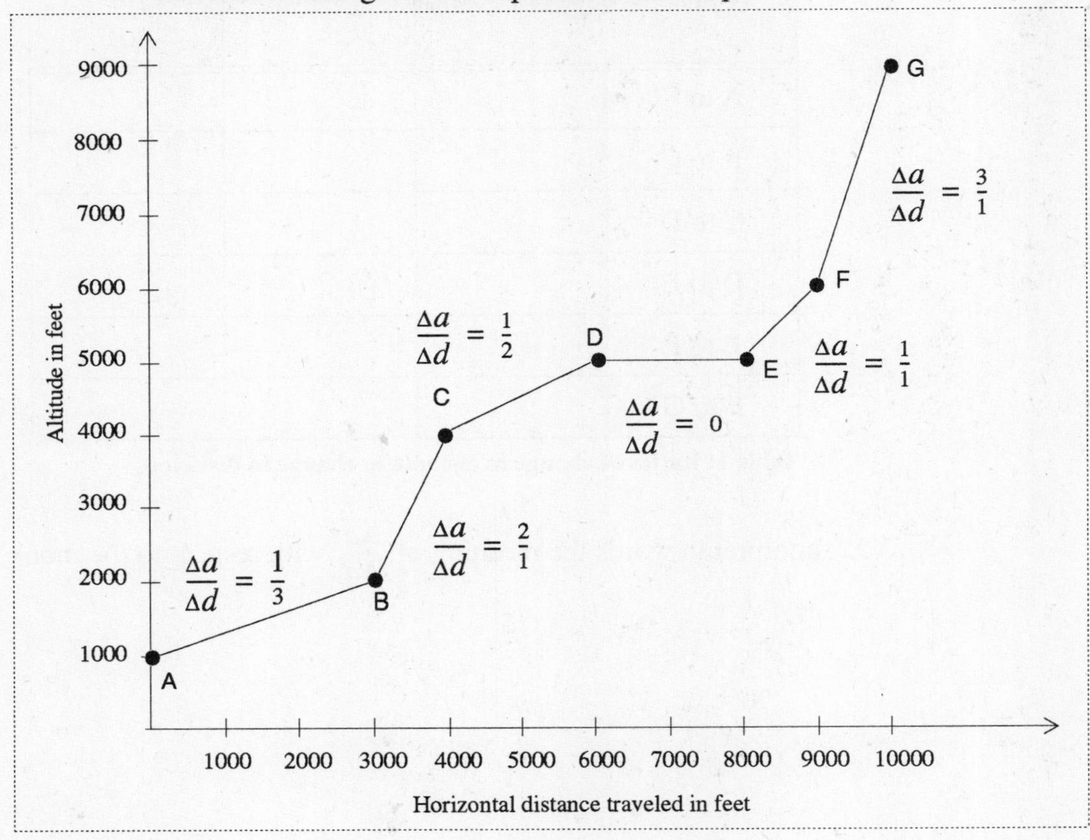

Figure 4

Each value of the ratio is called the ***slope*** of the line segment. In this case, we vary slopes depending on our position on the mountain.

Investigation

Recall the car rental problem: My car is in the shop so I must rent a car for only one day. If I rent the car for one day, the charge is \$35 plus 17¢ per mile.

9. If m represents the number of miles driven and C represents the total cost of the rental, express C as a function of m.

11. Complete Table 2 for the car rental relationship.

Point	Miles driven m	Total cost C
A	0	
B	25	
C	50	
D	75	
E	100	
F	125	
G	150	

Table 2: Car rental

12. Complete Figure 5. Find the finite differences for the miles column and the total cost column. Use the values from Table 2.

Pt	Miles driven m	Finite Difference Δm	Total cost C	Finite Difference ΔC
A				
B				
C				
D				
E				
F				
G				

Figure 5

13. Complete Table 3 by finding the ratio $\frac{\Delta C}{\Delta m}$ for each finite difference in Figure 5.

Endpoints	Δm	ΔC	$\frac{\Delta C}{\Delta m}$
A to B			
B to C			
C to D			
D to E			
E to F			
F to G			

Table 3: Ratios of change in total cost to change in miles driven

13. Interpret in words the meaning of $\frac{\Delta C}{\Delta m}$ with respect to the car rental.

14. Look at the equation you wrote in Investigation 9. Where does the slope $\frac{\Delta C}{\Delta m}$ appear in this equation?

Discussion

In the car rental problem, we find that the slope $\frac{\Delta C}{\Delta m}$ doesn't change and is therefore constant. Regardless of what two ordered pairs we choose, the ratio of change in cost to change in miles driven remains 0.17.

Since the algebraic relationship is $C(m) = 0.17m + 35$, we see that the slope appears as the number multiplying the variable m in the function. To see why the slopes are constant let's graph the values in Table 2. (See Figure 6)

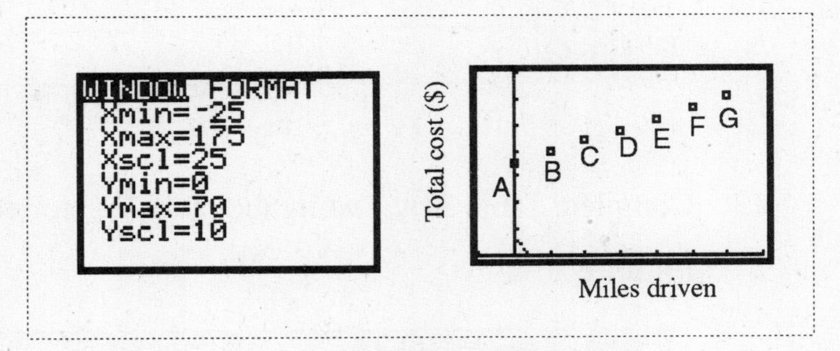

Figure 6

Notice that the points lie in a straight line indicating that the rate of change of total cost with respect to miles driven is constant.

In conclusion, we note that the rate of change or slope in both problems investigated in this section generally measures the ratio

$$\frac{\text{change in output}}{\text{change in input}}$$

We will visit this important idea again later in the text. For now, it's time to do some explorations. Enjoy.

Explorations

1. List and define words in this section that appear in ***italics bold*** type using your own words.

2. Recall the problem: General admission tickets to a concert sell for $19 each. The total receipts depend on the number of tickets sold.

    a. Create a table of at least five input/output pairs. Your choice of inputs should have constant finite differences (equal increments).

    b. Write the algebraic representation of the problem situation.

    c. Find the slope between any two input/output pairs.

    d. Describe in words the rate of change that the slope represents in this problem situation.

    e. What happens if the inputs are not in equal increments?

3. People with disabilities can enjoy national park trails thanks to the Uniform Federal Accessibility Standards which were established to inform hikers of trails' difficulty in terms of slope. A slope of no more than 1 foot in height for every 20 feet in distance is classified as Fully Accessible. A slope of no more than 1 foot in height for every 8 feet in distance is classified as Moderately Accessible. A slope of no more than 1 foot in height for every 6 feet in distance is classified as Significant Barriers.

    Classify each segment of the monk's path in Figure 1 on page 238.

4. Just the other day I was driving my gas guzzler (I have to get rid of that thing). I found that I was getting about 17 miles per gallon when I drove 45 miles per hour and 14 miles per gallon when I drove 60 miles per hour. This relationship between speed and gas mileage applies for speeds between 40 m.p.h. and 75 m.p.h.

    a. Find the average rate of change of the gas mileage with respect to the speed. Write a complete sentence describing your answer.

    b. Assuming that the rate of change in part a is constant, what is my gas mileage if I travel 50 miles per hour? What is my gas mileage if I travel 70 miles per hour. Defend your answers.

    c. What does the negative rate of change mean in this problem?

5. Recall the falling stale corned beef sandwich. Table 4 displays the data.

Time (seconds)	Distance fallen (feet)
1	16
2	64
3	144
4	256
5	400
6	576
7	784
8	1024
9	1296
10	1600

Table 4: The travels of a stale corned beef sandwich

a. For each one second interval from Table 4, calculate the change in the height. Record your answers in Table 5.

Change in time (seconds)	Change in distance fallen (feet)
1 to 2	
2 to 3	
3 to 4	
4 to 5	
5 to 6	
6 to 7	
7 to 8	
8 to 9	
9 to 10	

Table 5: Change in distance fallen of the stale corned beef sandwich

b. Is there a constant rate of change in height with respect to time or a varying rate of change in height with respect to time? Justify your answer.

c. In words, what does the average rate of change in height with respect to time represent?

d. Use the first and last table values from Table 4 to find the average rate of change in height with respect to time over the entire time period. Interpret the answer in terms of what is happening to the poor stale corned beef sandwich.

e. Over what one second interval is the sandwich moving the fastest? Over what one second interval is the sandwich moving the slowest? Justify your answers.

6. It was a beautiful day so I decided to take a long drive and clear my mind. The odometer on my gas guzzler (see Exploration 4) read 78,239 when I started the trip at 10 a.m. I stopped for lunch at noon and read the odometer which was at 78,334. I continued my trip at 1:15 p.m. and stopped to smell the roses at 2:30 with the odometer at 78,383. I left at 2:35 and arrived home at 5:10. The odometer then read 78,489.

   a. For each segment of the trip, find the rate of change of distance with respect to time.

   b. Find the average rate of change in distance with respect to time for the entire trip. Write a complete sentence describing your answer.

7. Refer to Figure 4 on page 242. The letters A and *a* are both used. Describe the difference in how these letters are used.

8. Slope is often represented as the ratio of the change in output to the change in input. Given any two points on a line, we can calculate the slope of the line by finding this ratio. More formally, if (x_1, y_1) and (x_2, y_2) are two points on a line then

$$slope = \frac{y_2 - y_1}{x_2 - x_1}.$$

   Draw the line and find the slope of the line given two points on the line.

   a. (3, 11) and (6, 4)

   b. (–5, 3) and (4, –6)

   c. (0, –8) and (–7, –4)

   d. (2, –3) an d (15, –3)

   e. (4, 2) and (4, –3)

9. Luis was hit on the head by a falling stale corned beef sandwich. (Where did that come from?) He angrily threw the sandwich as far as he could. The graph in Figure 7 displays the motion of the sandwich.

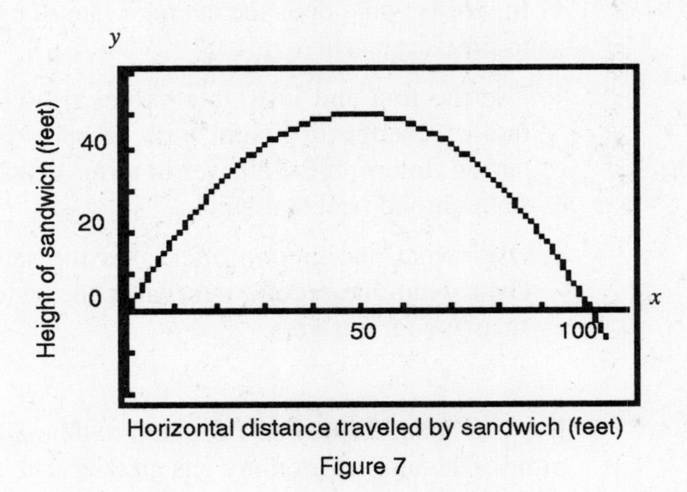

Figure 7

a. Mark and label the points on the graph beginning with a horizontal distance of 0 at increments of 20 feet horizontally.

b. For each consecutive pair of points labeled, find the rate of change of the height with respect to the horizontal distance traveled. Interpret each of your answers in terms of the motion of the sandwich.

c. What does it mean when the rate of change computed in part b is positive? What does it mean when the rate of change computed in part b is negative?

d. Approximate the ordered pair where the sandwich is at its highest point.

e. What is the rate of change of the altitude with respect to the horizontal distance traveled for the points nearest the highest point of the sandwich?

f. How does the rate of change in part e compare to those before the sandwich reaches its highest point? How does the rate of change in part e compare to those after the sandwich reaches its highest point?

Concept Map

Construct a concept map centered around the word **slope**.

Reflection

Use your concept map to write a paragraph describing the idea of slope in your own words. Describe at least two examples where you have encountered rate of change in your own life.

Section 5.3 Reciprocal and Power Functions

Purpose

- Investigate an application of a reciprocal function.
- Formalize the unary operation of reciprocaling as a function.
- Revisit power functions.
- Investigate the graphs of the reciprocal and power functions.

Investigation

Over spring break you want to visit a good friend who attends another college 240 miles away (wherever that is). The following graph displays information about the relationship between the average speed and the time required to make a 240–mile trip.

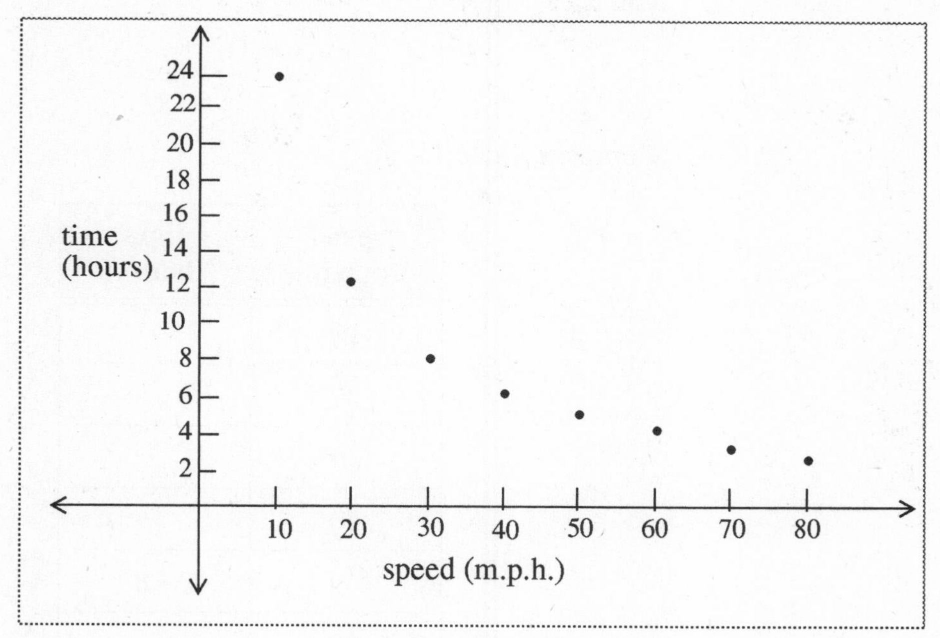

Figure 1

Use Figure 1 to answer the following questions.

1. As the average speed increases, does the time it takes to complete the trip increase or decrease? How do you know

   a. based on experience?

   b. based on the graph?

2. How long will the trip take if the average speed is 60 m.p.h.? What was the mode of transportation used?

3. You promised to be there, but your car broke down. So you took your bike. How long will the trip take if you're average speed is 10 m.p.h.?

4. You decide to make the trip by bus. How long will the trip take if the bus has an average speed of 45 m.p.h.?

5. You are running late! You promised you'd be there in one and one–half hours. How fast must you travel? What mode of transportation should you use?

6. Complete Table 1.

speed (m.p.h.)	time (hours)
10	
20	
30	
40	
	4.8
60	
	3.4
	3

Table 1: Speed versus time on a 240–mile trip

7. Does Table 1 agree with your answer to Investigation 1? Why?

Discussion

Figure 1 shows a relationship between average speed and time. We read a graph from left to right. To discuss the relationship, we scan the horizontal ***axis*** (input) from left to right. We always look at the input as it increases to interpret the change in the output.

- As the input increases, the output may increase, or
- As the input increases, the output may decrease, or
- As the input increases, the output may stay the same.
- Sometimes the output may change from increasing to decreasing or vice versa.

Investigation

8. In Figure 2, state the relationship between input and output for each graph.

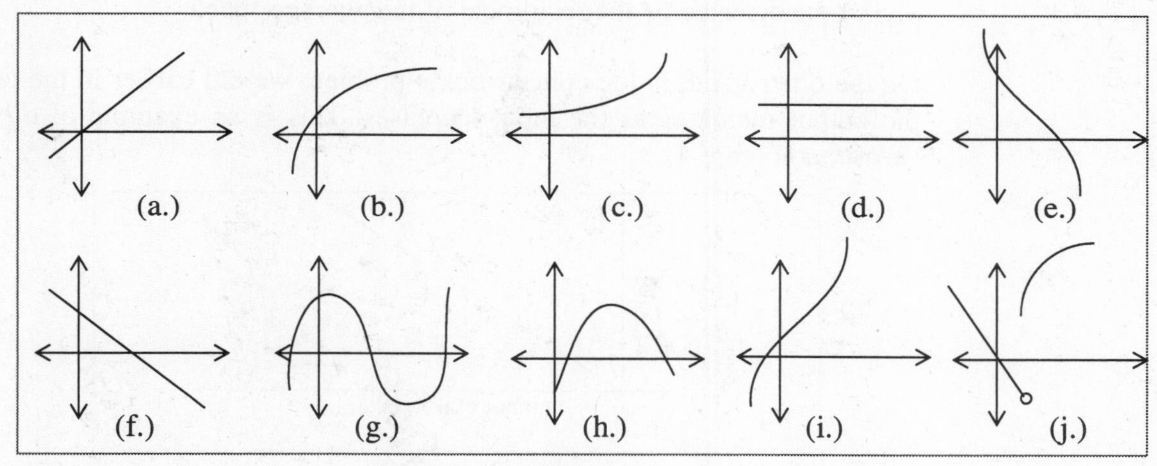

Figure 2

a. b.

c. d.

e. f.

g. h.

i. j.

9. For which graphs are your answers to Investigation 8 the same? How are these graphs the same? How are they different?

Discussion

When we look at the problem about the 240–mile trip, we see a graph in which the output (time) decreases as the input (average speed) increases where the product of the input and output is constant. This is an example of ***inverse variation*** (Figure 3).

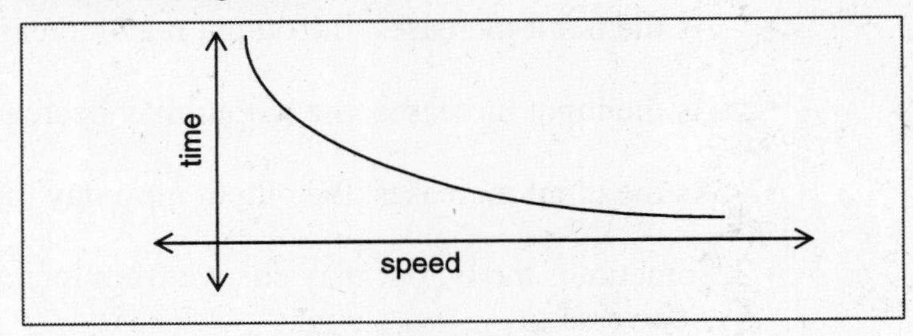

Figure 3

In general, two quantities ***vary inversely*** if their product is a constant.

Inverse variation involves a reciprocal function. In this example, distance is constant and we compute the time by dividing the distance by the average speed. From our understanding of division, this is equivalent to multiplying the distance by the reciprocal of the average speed.

On the other hand, in the concert ticket problem we did earlier in the text, the output increases as the input increases. This is an example of ***direct variation*** (Figure 4).

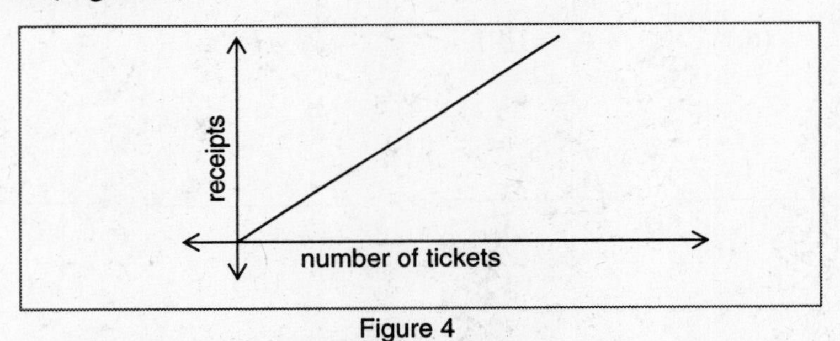

Figure 4

In general, two quantities ***vary directly*** if their ratio is a constant.

There are several other types of variation displayed in Figures 5–7.

Accelerated

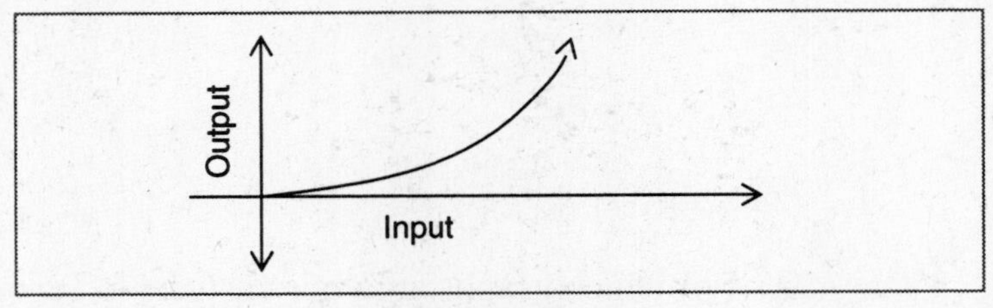

Figure 5

Cyclic

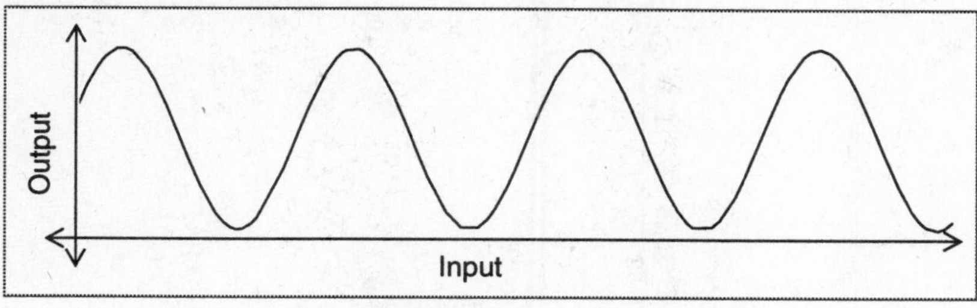

Figure 6

Stepped

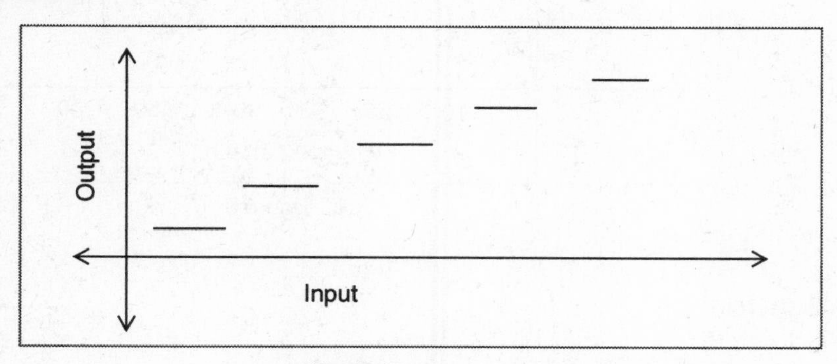

Figure 7

Each of these are used to represent relationships between changing quantities. While there are others, these are the most common.

Let's return to the relationship between average speed and the time required to make a 240–mile trip described at the beginning of this section. On a graph, we use a ***point*** to represent a correspondence between two numbers. For example, if the average speed is 40 m.p.h, the time for the trip is 6 hours.This pair of numbers is usually represented as an ***ordered pair***, written (40, 6). The input, 40, is written first and the second number, 6, in the ordered pair represents output. The value of the input is found along the horizontal axis. The value of the output is found along the vertical axis.

In this example, we have the ordered pair (speed, time). The pair of numbers (40, 6) is displayed by the point labeled A in Figure 8.

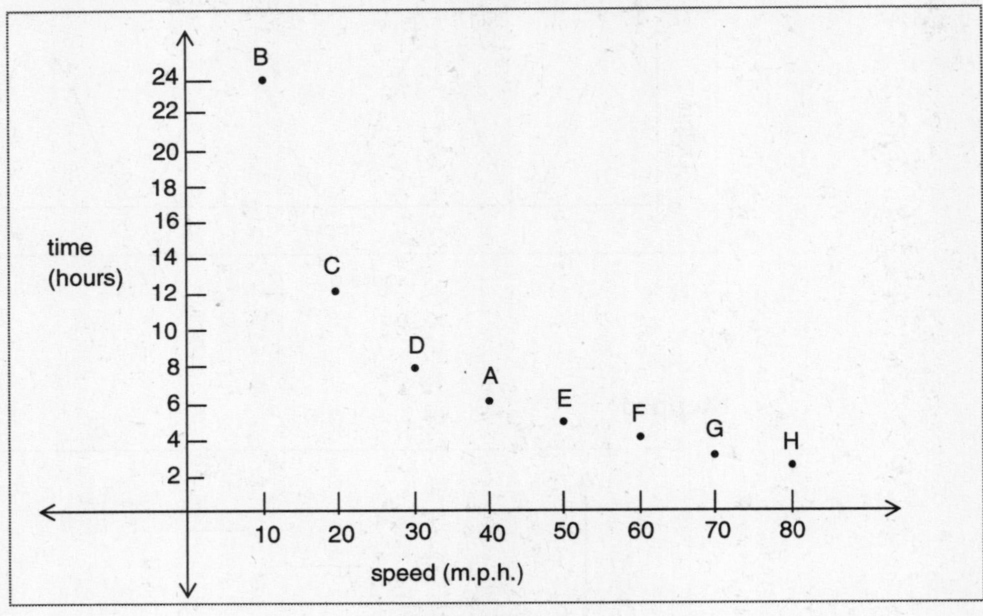

Figure 8

Investigation

10. Write the ordered pairs for each of the labeled points in Figure 8. Compare your answers to Table 1 on page 250.

11. Given the relationship between time and speed for a 240–mile trip, let r represent the average speed (in m.p.h.) and t represent the time (in hours). (Note: since speed is the rate of change of distance with respect to time, it is thought of as a rate and often represented by the variable r.)

    a. Express time t as a function of speed r.

    b. Draw a function machine for the function you wrote in part a.

12. Complete Table 2.

Number	Reciprocal of number
7	
–8	
$2\frac{3}{5}$	
$-\frac{7}{3}$	
–8.962	
5.37	
0	
x	

Table 2: Some reciprocals

13. Let $Rec(x)$ represent the function with input x and output the reciprocal of x. Write an algebraic representation (equation) for $Rec(x)$.

14. Use your calculator to generate five ordered pairs of $Rec(x)$. Include both positive and negative numbers for input.

15. a. Construct a graph with input x and output $Rec(x)$. Plot the points corresponding to the table created in Investigation 14.

b. Graph the $Rec(x)$ function on your graphing utility.

16. Earlier in the text, we studied power functions. Write an algebraic representation for

    a. the squaring function.

    b. the cubing function.

    c. the fourth–power function.

    d. the fifth–power function.

17. On your graphing utility graph each of the functions from Investigation 16.

Discussion

We have looked at one new function *Rec*(*x*) in the preceding investigations.

Table 2 and Investigation 14 illustrate the numeric representation of this function.

The algebraic representation is

$$Rec\,(x) \;=\; \frac{1}{x}.$$

The geometric representation appears in Figure 9.

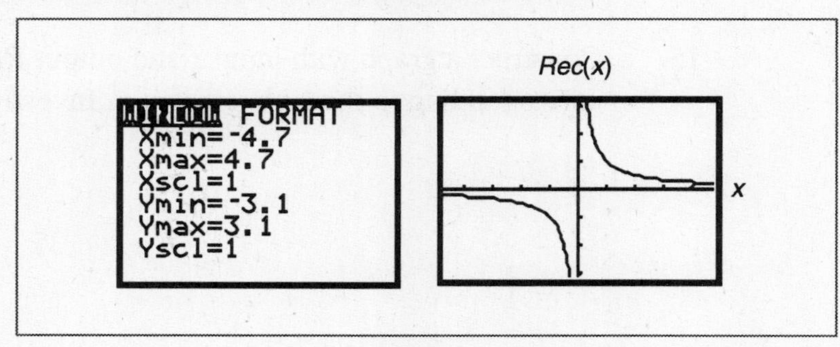

Figure 9

We have looked at the algebraic and geometric representations of some power functions. The equations and graphs follow.

- Squaring function $S(x) = x^2$. See Figure 10.

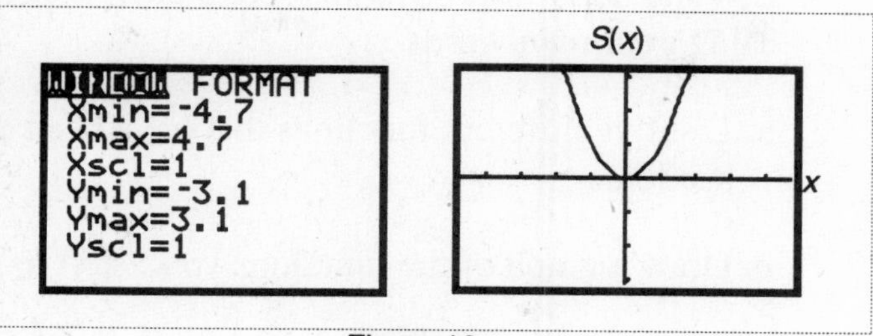

Figure 10

- Cubing function $C(x) = x^3$. See Figure 11.

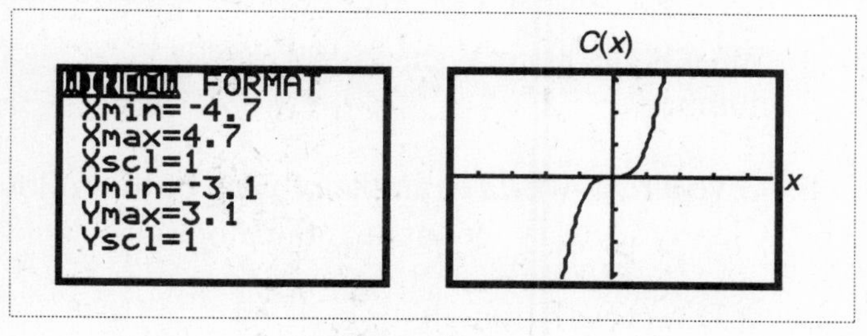

Figure 11

- The fourth–power function $Fourth(x) = x^4$. See Figure 12.

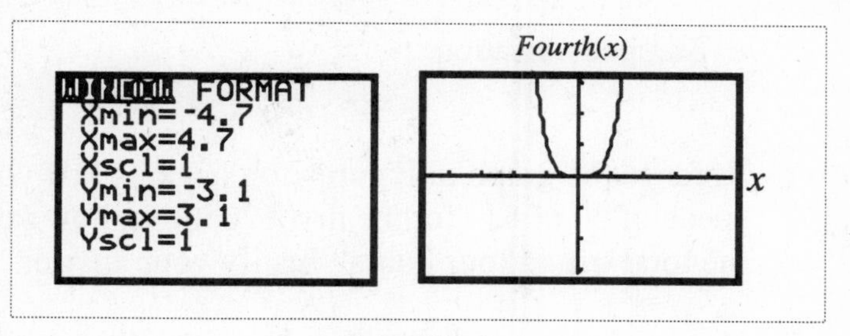

Figure 12

- The fifth–power function $Fifth(x) = x^5$. See Figure 13.

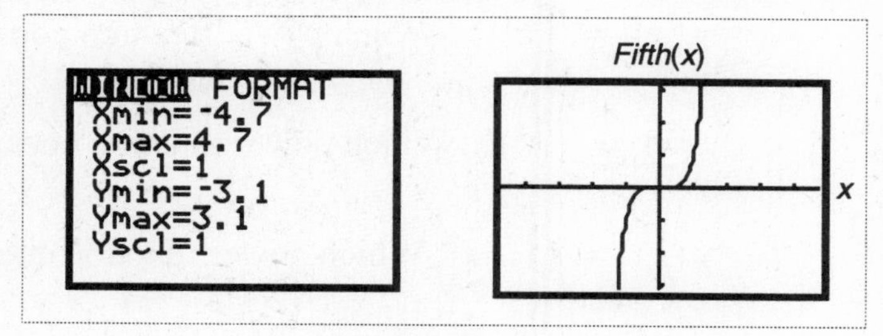

Figure 13

Explorations

1. List and define words in this section that appear in ***italics bold*** type using your own words.

2. a. List five different functions that have been discussed prior to this section.

   b. Draw a graph of the functions you listed in part a.

3. What is the general shape of $Power\,(x) = x^n$ if n is an odd whole number?

4. What is the general shape of $Power\,(x) = x^n$ if n is an even whole number?

5. Give a real–world example of each of the following types of variation. Clearly identify the input and the output.

   a. Direct variation.

   b. Indirect variation.

   c. Accelerated variation.

   d. Cyclic variation.

   e. Stepped variation.

6. Refer to the graph in Figure 1 on page 249. If you travel at an average speed of 30 m.p.h. for the first 120 miles, how fast must you go to make the total trip in four hours? Justify your answer.

7. Use a graphing utility to graph:

   a. $y\,(x) = (x^2)\,x$. Which power function does this graph most resemble?

   b. $y\,(x) = (x^3)\,x$. Which power function does this graph most resemble?

   c. $y\,(x) = (x^4)\,x$. Which power function does this graph most resemble?

   d. $y\,(x) = (x^3)\,x^2$. Which power function does this graph most resemble?

8. What single power function would the graph of $y\,(x) = (x^m)\,x^n$ most resemble?

9. Use a graphing utility to graph:

   a. $y(x) = (x^1)^3$. Which power function does this graph most resemble?

   b. $y(x) = (x^2)^3$. Which power function does this graph most resemble?

   c. $y(x) = (x^2)^4$. Which power function does this graph most resemble?

   d. $y(x) = (x^3)^2$. Which power function does this graph most resemble?

10. What single power function would the graph of $y(x) = (x^m)^n$ most resemble?

11. Use a graphing utility to graph:

    a. $y(x) = \frac{x^2}{x}$. Which power function does this graph most resemble?

    b. $y(x) = \frac{x^3}{x}$. Which power function does this graph most resemble?

    c. $y(x) = \frac{x^4}{x}$. Which power function does this graph most resemble?

    d. $y(x) = \frac{x^3}{x^2}$. Which power function does this graph most resemble?

    e. $y(x) = \frac{x^4}{x^2}$. Which power function does this graph most resemble?

    f. $y(x) = \frac{x^5}{x^2}$. Which power function does this graph most resemble?

12. What single power function would the graph of $y(x) = \frac{x^m}{x^n}$ most resemble?

13. Two variables a and b ***vary inversely*** if their product is always the same. For example, if a collection of rectangles has an area of 10 square feet, the lengths and widths of the rectangles vary inversely.

    Let w represent the width and act as input and l represent the length and act as output.

width	length
1	
2	
4	
5	
10	
w	

Table 3: Width versus length of a rectangle with area 10 square feet

a. Complete Table 3.

b. Describe how to find the length if you know the width.

c. Draw a function machine for the relationship in Table 3.

d. Write an algebraic representation (equation) for $l(w)$.

e. Graph this function on a graphing utility.

f. Use TRACE to find three more input/output pairs.

g. Use TRACE to find the width of the rectangle when the length is 3.8 feet.

Concept Map

Construct a concept map centered on the word **variation**.

Reflection

Give a real–life example of two quantities that are inversely related. Identify the input. Identify the output.

Section 5.4 Integer Exponents

Purpose

- Collect data involving exponential expressions.
- Formulate generalizations about properties of integer exponents.

Investigation

Recall the S P C system and the fact that we generated valid paths beginning with the initial path S P. Let's investigate the P–counts if we successively apply Rule 1.

1. Complete Table 1 assuming we begin with S P.

Number of applications of Rule 1	P–count
0	1
1	
2	
3	
4	
5	
6	
7	

Table 1: Rule 1 and P–counts

2. If Rule 1 is applied 20 times, use exponents to express the P–count.

3. If Rule 1 is applied n times, use exponents to express the P–count.

4. If Rule 1 is applied 20 times and then Rule 3 is applied once, use exponents to express the P–count. What is your answer if Rule 3 is applied twice? What is your answer if Rule 3 is applied three times?

If you do not see a pattern, apply Rule 1 only a few times and answer the question.

Problem Solving Toolkit

Discussion

Using the results of Table 1, we see that the exponent on the 2 equals the number of times Rule 1 was applied. So if Rule 1 is applied 20 times, the P–count would be 2^{20}. Generalizing, if Rule 1 is applied n times, the P–count is 2^n.

If Rule 1 is applied 20 times and then Rule 3 is applied once, the P–count is $2^{20} - 3$. If Rule 3 is applied twice, the P–count is $2^{20} - 6$. Finally, if Rule 3 is applied three times, the P–count is $2^{20} - 9$. These last three cases show that the P–count is not always a power of 2.

We will investigate properties involving exponents in this section. An expression that contains an exponent is called an ***exponential expression***. The quantity that the exponent acts on is called the ***base***.

- a^n is an exponential expression.
- a is the base.
- n is the exponent.

If n is a whole number, a^n represents the product of n factors of a.

For example, $2^5 = 2 \cdot 2 \cdot 2 \cdot 2 \cdot 2$ –the product of five factors of 2.

Investigation

5. Calculate $5^3 5^2$, 5^6, 25^6, and 5^5. Which two answers are the same?

6. Use a symbol manipulator to record the answers to the following.

   a. $(x^7)\ (x^3)$ b. $a^5 a^9$

   c. $t^4 \cdot t$ d. $(b)\ (b^{17})$

e. y^9y^{13}

f. $c^{34} \cdot c^{45}$

g. x^4y^3

h. $(a^5)\ (b^9)$

i. $t^3 + t^2$

j. $b + b^{17}$

7. Using the results of Investigation 6, what is the answer to:

 a. $(y^6)\ (y^4)$

 b. $y^5 \cdot y$

 c. $y^m y^n$

8. Describe in words your conclusion to Investigation 7c.

9. Using the factor explanation of exponent, explain why your answers to Investigation 7a. and 7c. are correct.

10. What happened when you multiplied exponential expressions that had different bases?

11. What happened when you added, rather than multiplied, exponential expressions?

Discussion

We see from the preceding investigations that if we multiply exponential expressions with like bases, we retain the base and add the exponents.

Symbolically, $(a^m)(a^n) = a^{m+n}$.

Why is this true? a^m means the product of m factors of a and a^n means the product of n factors of a. The product results in the multiplication of $m + n$ factors of a. (Figure 1)

$$\underbrace{(a \cdot a \cdot a \cdot \ldots \cdot a)}_{m \text{ factors of } a}\ \underbrace{(a \cdot a \cdot a \cdot \ldots \cdot a)}_{n \text{ factors of } a} = \underbrace{a^{m+n}}_{m+n \text{ factors of } a}$$

Figure 1

For example, $(2^4)(2^3) = 2^{4+3} = 2^7 = 128$. Check it out on your calculator.

Verifying,

$$2^4 2^3 = (2 \cdot 2 \cdot 2 \cdot 2)(2 \cdot 2 \cdot 2) = 2 \cdot 2 \cdot 2 \cdot 2 \cdot 2 \cdot 2 \cdot 2 = 2^7$$

Two characteristics must be present to use this property.

- The bases in the two expressions must be the same.
- The operation on the expressions must be multiplication.

We can't simplify $(a^5)(b^4)$ because the bases are different.

We can't simplify $a^5 + a^4$ because the expressions are being added, not multiplied.

Investigation

12. Calculate $(3^4)^2$, 3^6, 9^4, and 3^8. Which answers are the same?

13. Use a symbol manipulator to record the answers to the following.

a. $(x^7)^3$ b. $(a^5)^9$

c. $(t^2)^4$ d. $(z^3)^7$

14. Using the results of Investigation 13, what is the answer to:

a. $(y^6)^4$ b. $(y^m)^n$

15. Describe in words your conclusion to Investigation 14b.

16. Using the factor explanation of exponent, explain why your answers to Investigation 14a. and 14b. are correct.

Discussion

If we raise an exponential expression to a power, we retain the base and multiply the exponents.

Symbolically, $(a^m)^n = a^{mn}$.

Why is this true? $(a^m)^n$ means the product of n factors of a^m. Since we are multiplying exponential expressions with like bases, we can add the exponents. This means that the base is a and the exponent is the sum of n m's. This exponent is the product of m and n. (See Figure 2).

$$(a^m)^n = \underbrace{a^m \cdot a^m \cdot a^m \cdot \ldots \cdot a^m}_{n \text{ factors of } a^m} = a^{\overbrace{m+m+m+\ldots+m}^{n\ m\text{'s in exponent}}} = a^{mn}$$

Figure 2

For example, $(7^3)^5 = 7^{(3)(5)} = 7^{15}$. Check it out on your calculator.

Verifying:

$$(7^3)^5 = 7^3 \cdot 7^3 \cdot 7^3 \cdot 7^3 \cdot 7^3 = 7^{3+3+3+3+3} = 7^{3 \cdot 5} = 7^{15}$$

Investigation

17. Use a symbol manipulator or data sheet to record the answers to the following.

a. 7^0

b. 11^0

c. 5^0

d. a^0

18. a. What is true about a^0?

b. Does your answer to part a hold if $a = 0$?

19. Use a symbol manipulator to record the answers to the following.

a. 2^{-1}

b. 3^{-1}

c. 5^{-1}

d. 2^{-3}

e. 3^{-2}

f. x^{-1}

g. y^{-2}

h. z^{-3}

20. Use the results of Investigation 19 to answer the following.

a. Rewrite 7^{-5} using a positive exponent.

b. Rewrite y^{-m} using a positive exponent. Assume that m is positive.

Discussion

From Investigation 17, we see that a non–zero base raised to the zero power is 1.

Symbolically, $a^0 = 1$ if $a \neq 0$.

We can interpret a^0 as representing no factors of a. In every product a factor of 1 is understood, so a^0 is 1 with no factors of a. This is demonstrated in Figure 3.

$$a^4 = 1aaaa$$
$$a^3 = 1aaa$$
$$a^2 = 1aa$$
$$a^1 = 1a$$
$$a^0 = 1$$

Figure 3

For example, $43^0 = 1$. Check it out on your calculator.

From Investigations 19 and 20, we see that a base raised to a negative exponent is equal to the reciprocal of the base raised to an exponent that is the opposite of the original exponent.

Symbolically, $a^{-n} = \frac{1}{a^n}$ providing that a is not zero. Why can't a be zero?

This property relies on making two changes.

- Replace the base a with its reciprocal $\frac{1}{a}$.
- Replace the exponent on a with the exponent's opposite.

For example, $2^{-3} = \frac{1}{2^3} = \frac{1}{8}$. Check it out on your calculator.

We can think of this property in terms of multiplicative inverses.

To do so, think of $-n$ as $(n)(-1)$.

Then $a^{-n} = (a^n)^{-1}$.

The right side represents the multiplicative inverse, or reciprocal, of a^n.

So $a^{-n} = (a^n)^{-1} = \frac{1}{a^n}$.

Investigation

21. Use a symbol manipulator to record the answers to the following.

a. $\frac{x^7}{x^3}$

b. $\frac{a^9}{a^2}$

c. $\frac{t^4}{t}$

d. $\frac{k^4}{k^7}$

e. $\frac{p}{p^4}$

f. $\frac{r^7}{r^7}$

g. $\frac{b^{39}}{b^{17}}$

h. $\frac{a^7}{b^4}$

22. Using the results of Investigation 21, what is the answer to:

a. $\frac{y^8}{y^3}$

b. $\frac{y^5}{y^7}$

c. $\frac{y^m}{y^n}$

23. Describe in words your conclusion to Investigation 22c.

24. Using the factor explanation of exponent, explain why your answers to Investigation 22a. and 22b. are correct.

25. What happened when we divided exponential expressions that had different bases?

Discussion

We see from the preceding investigations that if we divide exponential expressions with like bases, we retain the base and subtract the exponents.

Symbolically, $\frac{a^m}{a^n} = a^{m-n}$ provided that a is not zero. Why?

Let's see why this is true. The numerator contains the product of m factors of a and the denominator contains the product of n factors of a. If we reduce the fraction, we are left with $|m - n|$ factors of a. If m is larger than n, the remaining a's are in the numerator. If m is smaller than n, the remaining a's are in the denominator.

For example, $\frac{2^7}{2^3} = 2^{7-3} = 2^4 = 16$. Check it out on your calculator.

To verify, $\frac{2^7}{2^3} = \frac{2 \cdot 2 \cdot 2 \cdot 2 \cdot 2 \cdot 2 \cdot 2}{2 \cdot 2 \cdot 2} = \frac{2 \cdot 2 \cdot 2 \cdot 2}{1} = 2^4$

Also, $\frac{2^3}{2^7} = 2^{3-7} = 2^{-4} = \frac{1}{2^4} = \frac{1}{16}$

To verify,

$$\frac{2^3}{2^7} = \frac{2 \cdot 2 \cdot 2}{2 \cdot 2 \cdot 2 \cdot 2 \cdot 2 \cdot 2 \cdot 2} = \frac{1}{2 \cdot 2 \cdot 2 \cdot 2} = \frac{1}{2^4}$$

Two characteristics must be present to use this property.

- The bases in the two expressions must be the same.
- The operation on the expressions must be division.

We can't simplify $\frac{a^5}{b^4}$ because the bases are different.

Investigation

26. Calculate $(3^4 4^3)^2$, $3^8 4^6$, $3^6 4^5$, 12^9 and 12^{24}. Which answers are the same?

27. Use a symbol manipulator to record the answers to the following.

 a. $(x^3 y^4)^5$ b. $(a^7 b^5)^2$

 c. $(cd^3)^4$ d. $(r^6 t)^3$

28. Using the results of Investigation 27, what is the answer to

 a. $(x^2 y^3)^4$ b. $(x^m y^n)^p$

29. Use a symbol manipulator to record the answers to the following.

 a. $\left(\frac{x^3}{y^4}\right)^5$ b. $\left(\frac{a^7}{b^5}\right)^2$

 c. $\left(\frac{c}{d^3}\right)^4$ d. $\left(\frac{r^6}{t}\right)^3$

30. Using the results of Investigation 29, what is the answer to:

 a. $\left(\frac{x^2}{y^3}\right)^4$ b. $\left(\frac{x^m}{y^n}\right)^p$

Discussion

We see from Investigations 27 and 28 that exponentiation distributes over a product.

Symbolically, $(a^m b^n)^p = (a^m)^p (b^n)^p = a^{mp} b^{np}$.

This is true since the commutative and associative properties for multiplication can be used to reorganize the product so that all factors with base a are grouped together and all factors with a base of b are grouped together. This is demonstrated in Figure 4.

$$(a^m b^n)^p = \underbrace{a^m b^n \cdot a^m b^n \cdot a^m b^n \cdot \ldots \cdot a^m b^n}_{p \text{ factors of} a^m b^n}$$

$$= \underbrace{(a^m \cdot a^m \cdot a^m \cdot \ldots \cdot a^m)}_{p \text{ factors of } a^m} \underbrace{(b^n \cdot b^n \cdot b^n \cdot \ldots \cdot b^n)}_{p \text{ factors of } b^n} = a^{mp} b^{np}$$

Figure 4

For example, $(2^4 3^5)^2 = (2^4 3^5)(2^4 3^5) = (2^4 2^4)(3^5 3^5) = 2^8 3^{10}$.

The problem can be done as follows too.

$(2^4 3^5)^2 = (2^4)^2 (3^5)^2 = (2^{(4)2})(3^{(5)2}) = (2^8)(3^{10}) = 2^8 3^{10}$.

Check it out on you calculator.

We see from Investigations 29 and 30 that exponentiation distributes over a quotient.

Symbolically, $\left(\frac{a^m}{b^n}\right)^p = \frac{(a^m)^p}{(b^n)^p} = \frac{a^{mp}}{b^{np}}$ if b is not zero. Why?

This is true because we multiply fractions by multiplying numerators to obtain the new numerator and multiply denominators to obtain the new denominator as demonstrated in Figure 5.

$$\left(\frac{a^m}{b^n}\right)^p = \underbrace{\frac{a^m}{b^n} \cdot \frac{a^m}{b^n} \cdot \frac{a^m}{b^n} \cdot \ldots \cdot \frac{a^m}{b^n}}_{p \text{ factors of } \frac{a^m}{b^n}} = \frac{\overbrace{a^m \cdot a^m \cdot a^m \cdot \ldots \cdot a^m}^{p \text{ factors of } a^m}}{\underbrace{b^n \cdot b^n \cdot b^n \cdot \ldots \cdot b^n}_{p \text{ factors of } b^n}} = \frac{(a^m)^p}{(b^n)^p} = \frac{a^{mp}}{b^{np}}$$

Figure 5

For example, $\left(\frac{2^4}{3^5}\right)^2 = \frac{2^4 2^4}{3^5 3^5} = \frac{2^4 2^4}{3^5 3^5} = \frac{2^8}{3^{10}}$.

The problem can be done as follows too.

$\left(\frac{2^4}{3^5}\right)^2 = \frac{(2^4)^2}{(3^5)^2} = \frac{2^{(4)2}}{3^{(5)2}} = \frac{2^8}{3^{10}}$. Check it out on your calculator.

Explorations

1. List and define words in this section that appear in ***italics bold*** type using your own words.

2. Given the exponential expression y^m,

   a. what is the base?

   b. what is the exponent?

3. Use exponential properties, where possible, to rewrite each of the following. If you can't rewrite, explain why.

   a. $(b^3)(b^{11})$ b. $x^7 + x^4$

   c. $(z^5)(a^6)$ d. $(k^8)k$

   e. Check your answers numerically by substitution if $a = 3$, $b = 3$, $x = 4$, $z = 2$, and $k = 4$.

4. Use an exponential property to simplify the following.

a. $(t^5)^3$

b. $(r^2)^9$

5. Use an exponential property to simplify the following.

a. 357^0

b. π^0

c. t^0 if $t \neq 0$

6. Use an exponential property to rewrite without a negative exponent.

a. 8^{-2}

b. 7^{-1}

c. k^{-5} if $k \neq 0$

7. Use exponential properties to simplify if possible. If not possible, state why. Leave no negative exponents in your answer.

a. $\frac{k^7}{k^2}$

b. $\frac{t^4}{t^{11}}$

c. $\frac{z^5}{z}$

d. $\frac{x}{x^3}$

e. $\frac{x^4}{y^2}$

f. $\frac{y^{17}}{y^{17}}$

8. Use exponential properties to simplify.

a. $(a^4b)^3$

b. $\left(\frac{x^3}{y^7}\right)^2$

9. Use exponential properties to simplify. Leave no negative exponents in your answers.

a. $(-3x^4y^2)(2xy^5)$

b. $(2a^5b^4)^3(ab^5)^2$

c. $(-x^2)^3$

d. $(-x^3y^5)(xy^7)(x^4y)$

e. $(-a^4b^7)^5(-a^7b^3)^2$

f. -5^2

g. $(-5)^2$

h. -5^{-2}

i. $(-5)^{-2}$

j. $(7^{-1})^{-1}$

k. $-(-11)^0$

l. $(3^{-4})^3$

m. $(x^2)^{-3}$

n. $x^{-7}y^{-5}$

o. $(xy^{-5})^{-3}$

p. $\dfrac{a^{-4}b^3}{a^{-7}b^4}$

q. $(x^{-3}y)(x^5y^2)$

r. $(a^4b^{-7})(a^{-3}b^{-2})$

s. $(5x^{-1}y^3)^{-2}(3^{-2}x^{-5}y^{-1})^{-2}$

t. $\left(\dfrac{x^3y^7}{xy^4}\right)^3$

u. $(x+y)^{-1}$

v. $x^{-1}+y^{-1}$

10. Describe the difference in how you compute -3^2 as compared to $(-3)^2$.

11. Describe the meaning of a negative exponent.

12. a. In Table 2, the exponential expression appears in the left column. Write the answer to the expression in the right column.

Exponential Expression	Answer
2^5	
2^4	
2^3	
2^2	
2^1	
2^0	
2^{-1}	
2^{-2}	

Table 2: Patterns in Exponential Expressions.

b. Describe the pattern that appears in the answer column. Specifically, what can be done to one answer to get the answer below it?

c. Does the pattern you discovered help to justify the definition of zero and negative exponents? Explain.

Concept Map

Construct a concept map centered on the phrase **exponent properties**.

Reflection

State and clearly describe each exponential property discovered in this section. Give an example of each.

Section 5.5 Making Connections: Is there anything Rational about Functions?

Purpose

- Reflect upon ideas explored in Chapter 5.
- Explore the connections among representations of rational numbers.

Investigation

In this section you will work outside the system to reflect upon the mathematics in Chapter 5: what you've done and how you've done it.

1. State the five most important ideas in this chapter. Why did you select each?

2. Identify all the mathematical concepts, processes and skills you used to investigate the problems in Chapter 5?

3. What has been common to all of the investigations which you have completed?

4. Select a key idea from this chapter. Write a paragraph explaining it to a confused best friend.

5. You have investigated many problems in this chapter.

   a. List your three favorite problems and tell why you selected each of them.

   b. Which problem did you think was the most difficult and why?

Discussion

There are a number of really important ideas which you might have listed including rational number, rates of change, slope, reciprocals, power functions, exponents, and algebraic expressions.

Concept Map

Construct a concept map centered around the word **ratio.**

Reflection

What are the various ways 1/2 and 1/3 can be represented?
Explain how being able to express fractions in multiple ways makes it easier to understand rational number. How does it make it harder?

Illustration

Draw a picture of **a mathematics teacher**.

A MATHEMATICS
TEACHER

Chapter 6

Real Numbers: Completing a Mathematical System

Section 6.1 Real Numbers and the Algebraic Extension

Purpose

- Introduce irrational numbers.
- Introduce the real number system.

Investigation

I just bought a new dog, a beautiful yellow labrador retriever. The breeder suggested I prepare a dog pen. What size should it be? The breeder suggested that the square footage equal the anticipated weight of the dog. The dog should grow to a weight of about 100 pounds. So I guess I need a dog pen that is 100 square feet.

1. Suppose the dog pen is in the shape of a square. How long is each side? How did you find the answer? Is your answer exact? How do you know?

2. Now I know I'm crazy. I just bought another labrador retriever. This one is black. I need to double the size of the dog pen or the original dog will be jealous. I still want the dog pen to be square.

   a. How many square feet do I need now?

b. How long is each side? How did you find your answer? Is your answer exact? How do you know?

3. I've decided to make a circular dog pen instead of a square dog pen. I would like the area to be 100 square feet. What value should I use for the radius of the circle? (Note: The area of a circle is found by multiplying π by the square of the radius.) How did you find the answer? Is the answer exact? How do you know?

Discussion

To build the original dog pen (100 square feet), we need the square root of 100 to find the length of the side. The square root of 100 is a number whose square is 100. This value is 10. I know it is exact since $10^2 = 100$.

However, when we double the size of the dog pen, we need an area of 200 square feet. The length of each side is the square root of 200. To represent this exactly, write the length of the side as $\sqrt{200}$ feet. In most cases we'd prefer a decimal estimate. This value is between 14 and 15 feet since $14^2 = 196$ and $15^2 = 225$. Regardless of how many decimal places we record, the square of the answer will not equal 200 exactly. There is no way to write an exact decimal representation of the square root of 200. A similar problem occurs when we try to make a circular pen with area 100 square feet. An exact value of the radius involves both π and a square root. Again a decimal estimate is preferred. The radius must be between 5 and 6 feet since $(5^2)\pi \approx 78.5$ and $(6^2)\pi \approx 113.1$. Again an exact decimal value cannot be written. Even an exact decimal for π cannot be written.

When we compute square roots of rational numbers, the outputs are often decimals. These answers are approximations of the actual square roots. In fact the square roots are represented by decimals that do not end and that do not repeat a sequence of digits. They are called nonterminating, nonrepeating decimals. These numbers are not rational—they are called ***irrational numbers***. When we write an approximation for a number, we use the symbol "≈" to indicate that the result is not exact.

For example, $\sqrt{2} \approx 1.414$.

Decimals come in three forms, two of which represent rational numbers and one that represents irrational numbers.

- ***Terminating***: these decimals have a finite number of non–zero digits. An example is 0.625 which equals the rational number $\frac{5}{8}$. Any terminating decimal is a rational number.

- ***Repeating***: these decimals repeat a sequence of digits infinitely. An example is 1.857142857142857142857142857142857... which equals the rational number $\frac{13}{7}$. Any repeating decimal is a rational number. The repeating sequence of digits 857142 represent the ***period*** of the decimal.

- ***Nonterminating, nonrepeating:*** these decimals have an infinite number of non–zero digits and do not repeat a sequence of digits. Any nonrepeating, nonterminating decimal is an irrational number. The first 100 digits in the approximation of $\sqrt{2}$ are

  1.4142135623730950488016887242096980785696718753769
  4807317667973799073247846210703885038753432764157 3

  If you want to see more digits you can use your symbol manipulator. The number $\sqrt{2}$ is not a rational number.

Be careful not to overgeneralize. A lack of a repetition of digits in the first 50 or 100 decimal places does not guaranteee a repetition never occurs. The rational number $\frac{1}{113}$ has 112 digits in its period.

Irrational numbers arise in many places in addition to the operation of finding roots. For example, the ancient Greeks spent a great amount of time investigating the relationship between the circumference and the diameter of a circle. They knew that the ratio of circumference to diameter was a little more than 3, but they were stymied in their quest to find the exact decimal value for this ratio. In some sense, their search was fruitless since this ratio is the irrational number we now know as π.

The first two hundred digits in the decimal approximation of π are

3.1415926535897932384626433832795028841971693993751058
209749445923078164062862089986280348253421170679821480
865132823066470938446095505822317253594081284811174502
84102701938521105559644622948954930382.

Pretty fascinating!

Intuitively it appears there is no repetition. There is no guarantee that a pattern will not occur later in the decimal representation. This is a problem with the calculator display of decimals. Since calculators can display only a small number of digits, it is often not clear whether a decimal is terminating, repeating, or nonterminating, nonrepeating. When we see a decimal on a calculator, we may not be able to tell if it represents a rational or an irrational number.

Any decimal that results from a rational number by dividing numerator by denominator will terminate or repeat.

Investigation

4. Using long division (not your calculator), obtain the exact decimal for

   a. $\frac{1}{13}$ b. $\frac{2}{13}$

5. How do you know when you can quit when you are using long division to find the exact decimal representation of a fraction?

6. Think about the remainders that occur as you use long division to obtain the exact decimal representation for a fraction.

   a. How many different remainders are possible if the denominator is 13? Why?

   b. For any rational number of the form $\frac{a}{b}$ where $b \neq 0$, explain why the decimal representation must either terminate or repeat.

Discussion

Using long division, we discover that

$$\frac{1}{13} = 0.\overline{076923} \text{ and } \frac{2}{13} = 0.\overline{153846}$$

As soon as we get a remainder that is a repeat of one that has occurred earlier in the division process, we can stop dividing. The sequence of digits will repeat from that point. There are exactly 13 possible remainders, 0 through 12, when the denominator is 13. If the remainder is zero, the decimal terminates. The decimal can have a period of at most 12 digits (since there are only 12 possible remainders) if the decimal is repeating.

In general, the decimal representation of the rational number $\frac{a}{b}$ where $b \neq 0$ can have at most $b - 1$ digits in its period if it is repeating. This demonstrates that the decimal representation for any rational number must terminate or repeat.

Now let's look at the square root operation in more detail. The square root of a whole number is irrational if the square root is not itself a whole number. It can be proven that both $\sqrt{2}$ and π are irrational numbers and that irrational numbers exist, but this is beyond the scope of this book.

Investigation

7. Why is the square root operation a unary operation?

8. Is the unary operation of finding a square root closed under the rational numbers? Justify your answer.

9. For each of the following, identify the decimal representation as terminating, repeating, or nonterminating and nonrepeating. Justify each answer.

   a. $\frac{7}{4}$

   b. $\frac{3}{11}$

   c. $\sqrt{6.25}$

   d. $\sqrt{44}$

   e. $\sqrt{\frac{16}{49}}$

   f. $\sqrt{17}$

   g. $\sqrt{121}$

   h. $\sqrt{\frac{4}{5}}$

   i. $\frac{1}{113}$

10. Classify each of the numbers in Investigation 5 as rational or irrational. Justify each answer.

11. The fraction $\frac{1}{17}$ is a rational number. Let's find the exact decimal representation.

    a. Write down the decimal representation from your calculator. Do not record the last digit since it may have been rounded.

    b. Look at the decimal representations of $\frac{2}{17}, \frac{3}{17}, \frac{4}{17}$, etc. Find the representation in which the last three digits of your answer to part a appear somewhere in the decimal. This will allow you to add more digits to your answer to part a. Write the new decimal. Remember: don't include the last digit.

c. Repeat part b until the decimal representation begins to repeat. At this point, you have the complete exact decimal representation of $\frac{1}{17}$.

Discussion

The unary operation of finding a square root is not closed under the rational numbers since the square root of many rational numbers is irrational. We need to extend our number system so that we have closure. The new number system should include all the rational numbers and all the irrational numbers. This new system is called the ***real number system***. The set of real numbers is symbolized by $\Re$.

The set of real numbers is the union of the set of rational numbers and the set of irrational numbers. We can picture the number systems we have studied as follows (Figure 1).

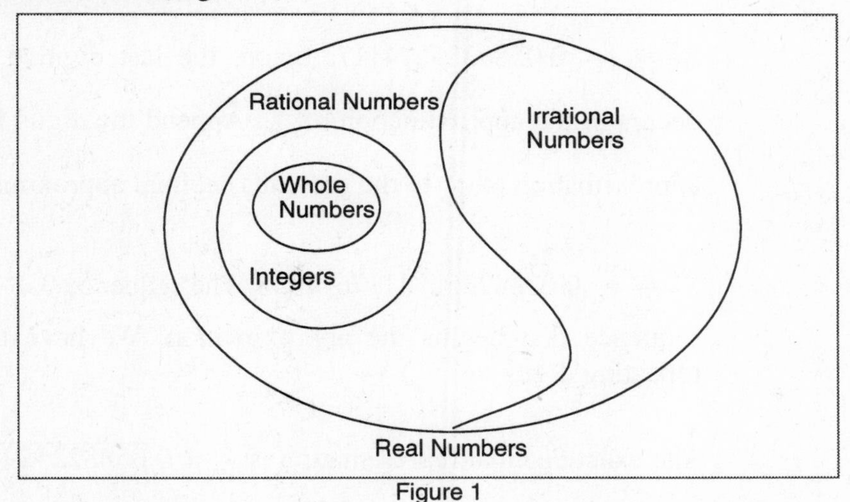

Figure 1

Note that this drawing is not to scale. There are actually many more irrational numbers than rational numbers.

All of the numbers in Investigation 9 are real numbers. Only $\sqrt{44}$, $\sqrt{17}$, and $\sqrt{\frac{4}{5}}$ are irrational numbers. The numbers $\sqrt{6.25}$, $\sqrt{\frac{16}{49}}$, and $\sqrt{121}$ are rational since their decimal representations terminate. The remaining numbers in the list are rational since they are the quotient of two integers.

The rational number $\frac{1}{17}$ has an exact decimal representation of $0.\overline{0588235294117647}$. The bar over the digits represents the sequence of digits that repeats. This decimal can be built by looking at the decimal representations of $\frac{1}{17}, \frac{2}{17}, \frac{3}{17}$, and $\frac{4}{17}$ (Figure 2).

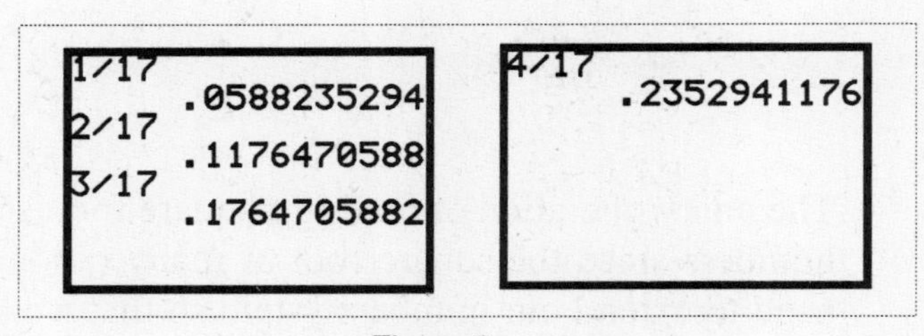

Figure 2

$\frac{1}{17} \approx 0.058823529$. Ignore the last digit, 4. The sequence 529 appear in the approximation of $\frac{4}{17}$. Append the digits following 529 in the approximation of $\frac{4}{17}$ to the decimal approximation of $\frac{1}{17}$.

So $\frac{1}{17} \approx 0.0588235294117$. Ignore the last digit, 6. The sequence 117 occurs in the approximation of $\frac{2}{17}$. Append the digits following 117 in the approximation of $\frac{2}{17}$ to the previous decimal approximation of $\frac{1}{17}$.

So $\frac{1}{17} \approx 0.058823529411764 7058$. The sequence 058 is a repetition of the sequence that begins the approximation. We have the complete set of repeating digits.

The exact decimal representation is $\frac{1}{17} = 0.\overline{0588235294117647}$.

The set of real numbers is in one–to–one correspondence with the set of points on a number line. This means

- every point on the number line has a real number that names it,
- every unique real number has a geometric representation—a point—on a number line.

This has profound implication for graphs of functions. A connected graph indicates that both the input and output variables may take on real number values, dependent on the problem situation. Let's investigate this further.

Investigation

12. In Section 4.3, we investigated a problem situation in which Okimbe spent $1000 on Chicago Bulls T–shirts and sold the shirts for $10 each. The function used the number of T–shirts sold as input and Okimbe's profit as output.

    a. Should the input variable be allowed to represent any real number value? Explain. If no, state the domain of the problem situation.

    b. Should the graph of the problem situation be connected or disconnected? Justify your answer.

13. In Section 5.3, we investigated the relationship between average speed and time required to make a 240–mile trip. The function used speed as input and time as output.

    a. Should the input variable be allowed to represent any real number value? Explain. If no, state the domain of the problem situation.

    b. Should the graph of the problem situation be connected or disconnected? Justify your answer.

Discussion

For Okimbe's T–shirt problem, the input represents the number of T–shirts he sells. This must be a whole number value. Therefore, the input variable is restricted to a whole number value. The graph should be disconnected. Such a problem situation is referred to as a ***discrete problem situation***. A good way to graph a discrete problem situation is to use parametric equations.

Recall that the original algebraic representation of the problem situation was $P(x) = 10x - 1000$. We represented this parametrically as

$$x = t$$

and

$$y = 10t - 1000.$$

First we set the calculator to parametric **and** dot mode. Enter the equations and define the viewing window by defining the setting on t. When we do this, we are enforcing the domain of the problem situation on the graph.

The smallest t can be is 0 so Tmin should be 0. The t values must be whole numbers so the setting for Tstep must be a whole number value. The setting of Tmax depends on the maximum number of T–shirts Okimbe will sell.

The settings for variable x are the same as those for t. The setting for y depend on the outputs of the function. The smallest y will be is -1000 which occurs if Okimbe sells no T–shirts. The largest y is determined by the largest number of T–shirts Okimbe sells.

Figure 3 displays one possible graph with window settings.

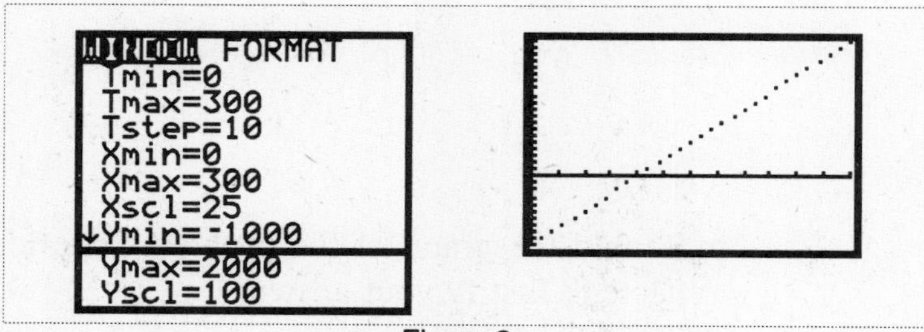

Figure 3

For the 240–mile trip, the input represents the speed. This is a measurement which can take on any nonnegative real number value. The graph should be connected. Such a problem situation is referred to as a ***continuous problem situation***. Continuous problem situations may be graphed using standard function mode. The algebraic representation of this problem situation is

$$t(r) = \frac{240}{r}$$

where r represents the speed in m.p.h. and t represents the time in hours. Set the calculator to function and connected mode and enter the function. For a viewing window, we might let the input value range from zero to eighty as values of speed. Let the output values (time) range from zero to thirty. Figure 4 displays the equation, viewing window, and graph.

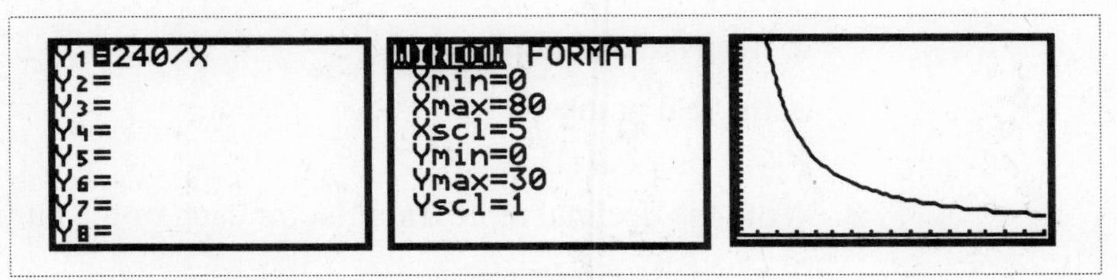

Figure 4

With the creation of the real numbers, we complete our excursion through the primary mathematical number systems. The real numbers are closed under the unary operations of oppositing, reciprocaling, and exponentiation, and square root extraction. The real numbers also are closed under the binary operations of addition, subtraction, multiplication, and division.

Unless stated otherwise, variables will represent real numbers.

In conclusion, we must note that there are still some other numbers to explore later in your mathematical career. Try finding the square root of a negative number on your calculator. What happens? We will need an additional set of numbers to deal with this problem.

Explorations

1. List and define the words in this section that appear in ***italics bold*** type.

Given the following collection of numbers:

3	$\lvert 2.3\rvert$	-7	$-\frac{2}{3}$	$\frac{3}{5}$
$\sqrt{7}$	$\lvert -3.1\rvert$	5	$-\sqrt{2}$	$\frac{13}{7}$
$-\frac{6}{11}$	$\sqrt{\frac{2}{3}}$	$-\sqrt{9}$	$\sqrt{5.72}$	0
$-\frac{15}{3}$	$-\lvert 4.6\rvert$	$\sqrt{\frac{9}{4}}$	$\sqrt{0.25}$	2^{-3}
1.3^2	$-\lvert -3.7\rvert$.			

2. List the numbers that belong to

   a. the whole numbers.

   b. the integers.

   c. the rational numbers.

   d. the irrational numbers.

   e. the real numbers.

3. Write the decimal representation for each of the numbers.

4. Place all of the numbers on a number line.

5. Rearrange the original list of numbers so they are written in order from the smallest number to the largest number. Justify your ordering.

6. For each number system, list the operations that are closed within the given number system.

   a. Whole numbers.

   b. Integers.

   c. Rational numbers.

   d. Real numbers.

7. Set a viewing window of Xmin = 0, Xmax = 1, Xscl = 0, Ymin = –1, Ymax = 1, and Yscl = 0 on your graphing utility.

   a. Trace along the x–axis. Are the x values rational or irrational numbers? How many points are on the x–axis between 0 and 1?

   b. Change Xmin to 0.1 and Xmax to 0.2. Trace along the x–axis. Are the x values rational or irrational numbers? How many points are on the x–axis between 0.1 and 0.2?

   c. Change Xmin to 0.01 and Xmax to 0.02. Trace along the x–axis. Are the x values rational or irrational numbers? How many points are on the x–axis between 0.01 and 0.02?

   d. Change Xmin to 0.001 and Xmax to 0.002. Trace along the x–axis. Are the x values rational or irrational numbers? How many points are on the x–axis between 0.001 and 0.002?

e. Change Xmin to 0.0001 and Xmax to 0.0002. Trace along the x–axis. Are the x values rational or irrational numbers? How many points are on the x–axis between 0.0001 and 0.0002?

f. Regardless of size of the interval on the x–axis, how many points are on the x–axis in the given interval?

8. Find the decimal representation of $\frac{1}{113}$. (Hint: Enter $\frac{x}{113}$ as a function on your calculator and display the table. Then use the technique discussed in this section to find the decimal representation of $\frac{1}{17}$.) How many digits are needed before the decimal begins repeating?

9. List the different values you have used in the past to approximate π. The Chinese used the ratio $\frac{355}{113}$ to approximate π.

   a. What is the best approximation?

   b. What criteria did you use to determine the "best" approximation?

   c. Why do we need an approximation for π?

   d. Are approximations of π always rational? Why or why not?

Concept Map

Construct a concept map centered on the phrase **real numbers**.

Reflection

Write a story problem that uses an irrational number either in the problem or in the solution.

Section 6.2 The Square Root Function

Purpose

- Formalize the unary operation of finding a square root as a function.
- Investigate the numeric, algebraic, and geometric representations of the square root function.
- Investigate several applications of the square root function.

Investigation

Remember the square dog pen we designed in Section 6.1. Given the area of the pen, we found the length of the sides by calculating the square root of the area. Let's investigate square roots in more depth.

1. Use calculator where necessary to complete Table 1.

Area of square (square units)	Length of the side of the square (units)
7 cm^2	
9 $ft.^2$	
0 sq. ft.	
5.37 sq. in.	
$\frac{2}{3}$ in^2	
1 sq. cm.	
x sq. m.	

Table 1: Finding the length of the side of a square given the square's area

2. Let $Sqrt(x)$ represent the function with input x and output the square root of x. Write an algebraic representation (equation) for $Sqrt(x)$.

3. Determine which is larger.

a. $(9)^2$ or $\sqrt{9}$

b. $\left(\frac{1}{2}\right)^2$ or $\sqrt{\frac{1}{2}}$

4. a. Is the square of a number a product or a factor?

   b. Is the square root of a number a product or a factor?

5. Compare $(-9)^2$ to $\sqrt{-9}$. Are both computations possible? Why or why not?

6. Consider the **squaring function**. Use your calculator to generate a table of input/output values. Use the table to answer the following questions.

   a. State the domain of the function; that is, the set of numbers that can be used as input. Justify your answer.

   b. State the range of the function; that is, the set of numbers that are output by the function. Justify your answer.

7. Consider the **square root function**. Use your calculator to generate a table of input/output values. Use the table to answer the following questions.

   a. State the domain of the function; that is, the set of numbers that can be used as input. Justify your answer.

   b. State the range of the function; that is, the set of numbers that are output by the function. Justify your answer.

8. Graph the squaring function and the square root function in the same viewing window. Write several sentences comparing and contrasting the graphs of the two functions.

Discussion

We have investigated the square root function in the preceding investigations.

Table 1 provides a numeric representation of this function. As we complete Table 1, we make decisions about the number of decimal digits to list. While sometimes arbitrary, the number of digits listed is often determined by the context of the problem and the importance of accuracy in the problem context.

The algebraic representation is $Sqrt(x) = \sqrt{x}$. The function machine appears in Figure 1.

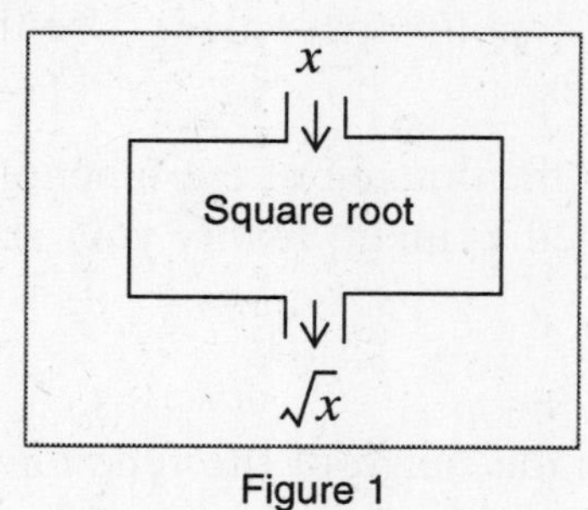

Figure 1

Be sure to differentiate between the squaring function and the square root function. We calculate squares by multiplying two identical factors. We find square roots by determining what number must be multiplied by itself to obtain the square.

The input to the squaring function is a factor and the output is the product. For example, if we input 9, the output is 81. Nine is a factor and 81 is the product of 9 times 9.

The input to the square root function is a product and the output is a factor, assuming the input was a square number. For example, if we input 9, the output is 3. Nine is the product of 3 and 3; three is the factor.

Figure 2 displays a table for the squaring function and square root function.

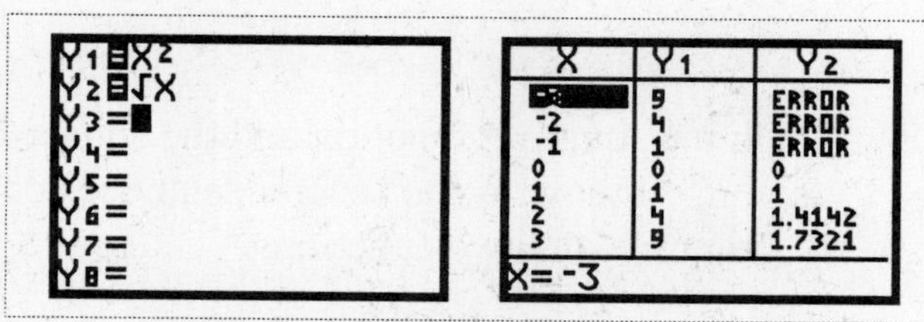

Figure 2

Figure 2 suggests that negative numbers may not be used as input to the square root function. In addition, the outputs of both the squaring function and the square root function are limited to non–negative numbers.

Figure 3 displays the geometric representation of the square root function.

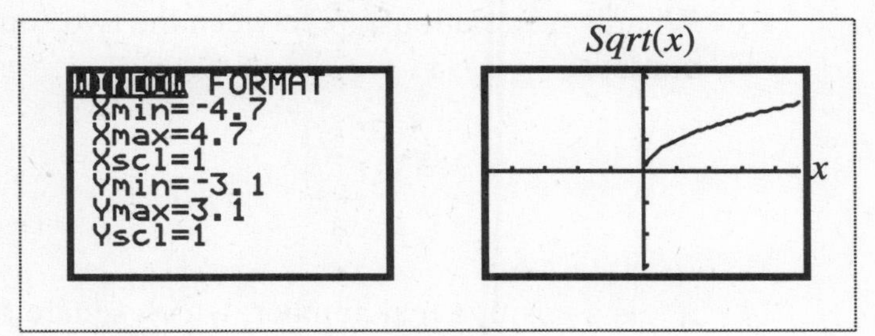

Figure 3

The set of legal inputs to a function is called the ***domain*** of the function. In this case, the inputs are limited to nonnegative real numbers. So the domain of $Sqrt\,(x)$ is the nonnegative real numbers.

The set of outputs of a function is called the ***range*** of the function. In this case, the output of *Sqrt*(x) is always a nonnegative real number. So the range of $Sqrt\,(x)$ is the nonnegative real numbers.

Investigation

9. a. There are two numbers you can square and get 9. What are they?

   b. There are two numbers you can square and get 36. What are they?

10. a. Use your calculator to find $\sqrt{9}$.

    b. Use your calculator to find $\sqrt{36}$.

    c. Use your calculator to find $-\sqrt{9}$.

    d. Use your calculator to find $-\sqrt{36}$

11. Are $-\sqrt{9}$ and $\sqrt{-9}$ different? Justify your answer.

12. How many square roots does a positive real number have?

13. What is the relationship between the two square roots of a positive real number?

14. If x is a positive real number, which square root of x do you get when you use $Sqrt(x) = \sqrt{x}$?

15. a. Graph $Sqrt(x) = \sqrt{x}$ on your graphing utility.

b. Graph $OppSqrt(x) = -\sqrt{x}$ on your graphing utility.

c. Graph $SqrtOpp(x) = \sqrt{-x}$ on your graphing utility.

c. How are the graphs of the function in parts a–c similar? How are they different?

Discussion

Every positive real number has two square roots. They are opposites of each other. For example, the two square roots of 25 are 5 and –5.

The positive square root is called the ***principal square root***. The ***principal square root*** of x is represented by $\sqrt{x}$.

The symbol √ is called a ***radical***. The x under the radical is called the ***radicand***. Historically, the letter "R" was placed in front of a number to indicate square root. The "R" gradually evolved into the radical symbol.

So $Sqrt(x) = \sqrt{x}$ only returns the nonnegative square root of x. The domain and the range are the set of nonnegative real numbers.

The negative square root of a nonnegative real number x is the opposite of the principal square root of x, symbolized by $-\sqrt{x}$. We can define the function opposite square root of x by

$$OppSqrt(x) = -\sqrt{x}.$$

The domain is the set of nonnegative real numbers and the range is the set of nonpositive real numbers.

The square root of the opposite of real number x is symbolized by $\sqrt{-x}$. We can define the function square root of the opposite of x by

$$SqrtOpp(x) = \sqrt{-x}.$$

The domain is the set of nonpositive real numbers and the range is the set of nonnegative real numbers.

Be sure to differentiate between $\sqrt{-x}$ and $-\sqrt{x}$. In the case of $\sqrt{-x}$, we take the opposite followed by finding the square root. Figure 4 displays the function machine and a graph in a standard viewing window.

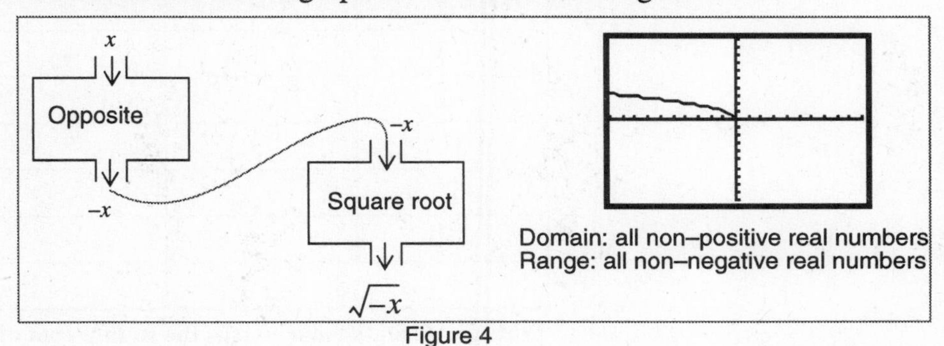

Figure 4

In the case of $-\sqrt{x}$, we take the square root followed by finding the opposite. Figure 5 displays the function machine.

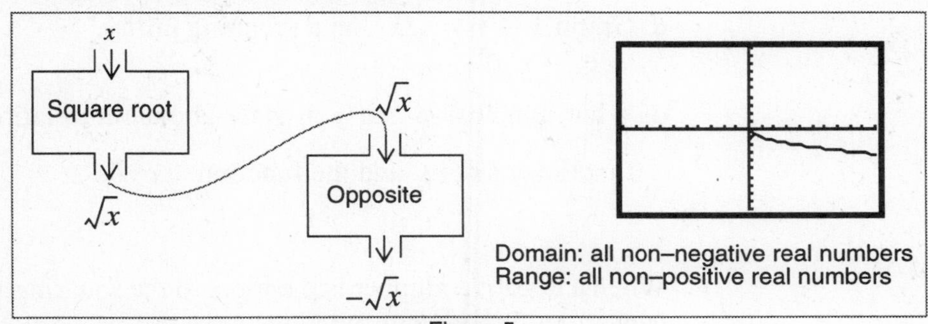

Figure 5

Explorations

1. List and define the words in this section that appear in ***italics bold*** type.

2. Use your calculator to evaluate each of the following. If the result is not a real number, say so and justify.

   a. $\sqrt{7}$ b. $-\sqrt{12}$

   c. $\sqrt{81}$ d. $-\sqrt{49}$

   e. $\sqrt{-49}$ f. $2\sqrt{11}$

   g. $-3\sqrt{5}$ h. $4\sqrt{9}$

   i. $3\sqrt{-2}$ j. $-7\sqrt{121}$

3. Construct a function machine for expressions i. and j. from Exploration 2.

4. a. Complete Table 2.

x	$abs(x)$	$\sqrt{x^2}$
−7		
−5		
−3		
−1		
0		
2		
4		
6		

Table 2: Absolute value versus the square root of a square

   b. Graph $y(x) = abs(x)$ on a graphing utility.

   c. Graph $y(x) = \sqrt{x^2}$ on a graphing utility.

   d. What conclusion can you state about the relationship between the function $abs(x)$ and the function $\sqrt{x^2}$?

5. When a negative number is the input to the squaring function the output is positive. What other functions have this property?

6. a. Complete Table 3. Write the first finite differences of the output in the third column.

x	x^2	$\Delta(x^2)$
1		
2		
3		
4		
5		

Table 3: Rate of change of the squaring function

b. Complete Table 4. Write the first finite differences of the output in the third column.

x	$\sqrt{x}$	$\Delta(\sqrt{x})$
1		
2		
3		
4		
5		

Table 4: Rate of change of the square root function

c. For parts a and b, are the outputs increasing or decreasing? Justify your answers.

d. Compare the finite differences of the outputs found in parts a and b. Which outputs are changing more quickly? Which outputs are changing more slowly? Justify your answers.

7. a. Complete Table 5

x	$\sqrt{x}$	$x^{1/2}$
1		
2		
3		
4		
5		
6		

Table 5: Comparing square roots with raising a number to the one–half power.

b. On a graphing utility, graph $y(x) = \sqrt{x}$.

c. On a graphing utility, graph $y(x) = x^{1/2}$. Be sure to enter the exponent in parentheses.

d. What is the relationship between $\sqrt{x}$ and $x^{1/2}$?

e. Use laws of exponents to justify why $\sqrt{x}$ and $x^{1/2}$ are different notations for the same function.

8. Calculate the following. If you use the exponential notation on you calculator, be sure to put the exponent in parentheses.

a. $3^{1/2}$

b. $16^{1/2}$

c. $100^{1/2}$

d. $17^{1/2}$

e. $-5^{1/2}$

f. $-49^{1/2}$

g. $(-4)^{1/2}$

9. Given two ordered pairs (a, b) and (c, d) representing two points in the plane, the distance between the two points can be found by

$$\sqrt{(\Delta x)^2 + (\Delta y)^2}$$

where Δx represents the change in the input and Δy represents the change in the output.

On your calculator, enter the ordered pairs and construct the line segment. Use the formula to find the length of each line segment. Compare the length of the line segment if the segment was placed along the x–axis. Does the length obtained by formula look correct?

a. (3, 8) and (5, 2)

b. (–6, 3) and (4, –9)

c. (–3, –5) and (–7, 2)

d. (8, –5) and (2, –5)

e. (4, 3) and (4, –6)

10. If an object is dropped from a height, the time t (in seconds) it will take the object to fall d feet is found by the function

$$t(d) = \sqrt{\frac{d}{16}}.$$

a. How long will it take the object to fall 30 feet?

b. How long will it take the object to fall one mile if it is dropped from an airplane?

c. A skydiver jumps from an airplane that is 9,500 feet high. She must open her chute when she is 3,000 feet above the ground. How long will she be in free fall?

11. When a driver slams on the brakes and the car skids to a stop, the minimum speed can be estimated by the function

$$S(d) = \sqrt{30fd}$$

where S is the speed of the car in m.p.h., f is the drag factor, and d is the length of the skid marks measured in feet. For this problem assume that $f = 0.83$. The drag factor takes into account conditions such as the type of pavement, whether the pavement is wet or dry, the conditions of the tires, etc.

a. What was the minimum speed of the car if the skid marks are 100 feet long?

b. A driver claims he was going the speed limit of 25 m.p.h.when he applied the brakes. The police measured the skid marks at the accident sight and found them to be 60 feet long. Should the driver be ticketed for speeding? Justify your answer.

c. If you are traveling at a rate of 65 m.p.h. and hit the brakes, how far will you skid before you come to a stop?

Concept Map

Construct a concept map centered on the phrase **square root**.

Reflection

Use your concept map as an aid to describe everything you learned about square roots while exploring this section.

Section 6.3 Classes of Basic Functions

Purpose

- Investigate the numerical, algebraic, and graphical representations of some basic functions.
- Identify the domain and range of some basic functions.
- Identify the intercepts of some basic functions.

Discussion

Throughout this text, we have looked at some important basic relationships. Table 1 includes the important basic functions that we have seen.

Function name	Basic form
Constant	$C(x) = a$, a a constant
Linear (Identity)	$L(x) = x$
Quadratic	$Q(x) = x^2$
Opposite	$Opp(x) = -x$
Absolute value	$Abs(x) = \|x\|$
Reciprocal	$Rec(x) = \frac{1}{x}$
Square root	$Sqrt(x) = \sqrt{x}$

Table 1: Basic functions and their algebraic representations

Investigation

1. a. Complete Table 2 for the constant function $C(x) = 3$.

Table 2: Constant function

Input	-3	-2	-1	$-\frac{1}{2}$	$-\frac{1}{4}$	0	$\frac{1}{4}$	$\frac{1}{2}$	1	2	3
Output											

b. Sketch the graph of the function. Clearly label your graph with the ordered pairs from the table.

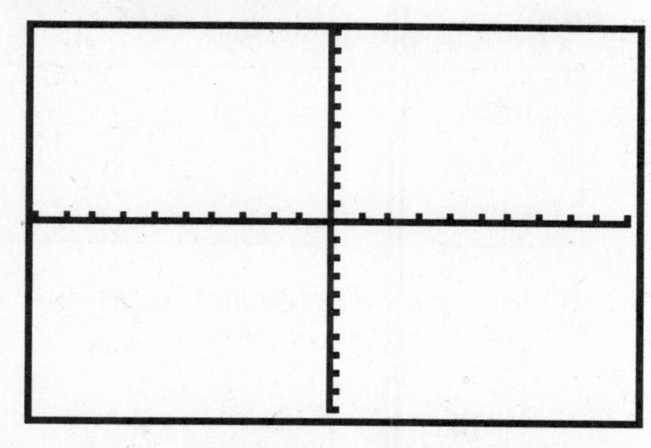

c. Compare and contrast the information you get from the table versus the equation versus the graph.

2. a. Complete Table 3 for the basic linear function $L(x) = x$.

Table 3: Linear function

Input	–3	–2	–1	$-\frac{1}{2}$	$-\frac{1}{4}$	0	$\frac{1}{4}$	$\frac{1}{2}$	1	2	3
Output											

b. Write the ordered pairs from Table 3 where the input is a whole number.

c. Sketch the graph of the function. Clearly label your graph with the ordered pairs from the table.

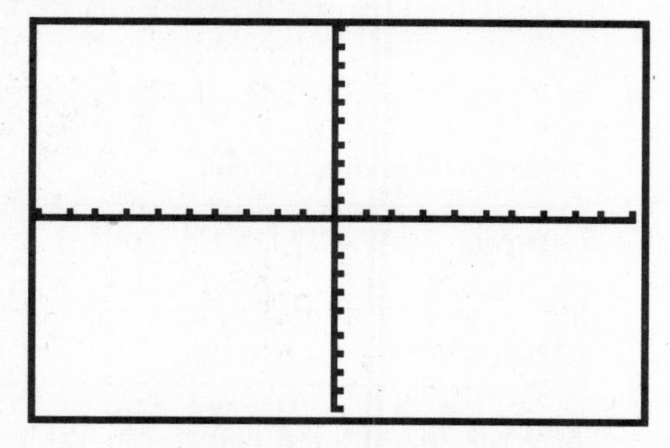

d. Compare and contrast the information you get from the table versus the equation versus the graph.

3. a. Complete Table 4 for the basic quadratic function $Q(x) = x^2$.

Table 4: Quadratic function

Input	–3	–2	–1	$-\frac{1}{2}$	$-\frac{1}{4}$	0	$\frac{1}{4}$	$\frac{1}{2}$	1	2	3
Output											

b. Write the ordered pairs from Table 4 where the output is 1.

c. Sketch the graph of the function. Clearly label your graph with the ordered pairs from the table.

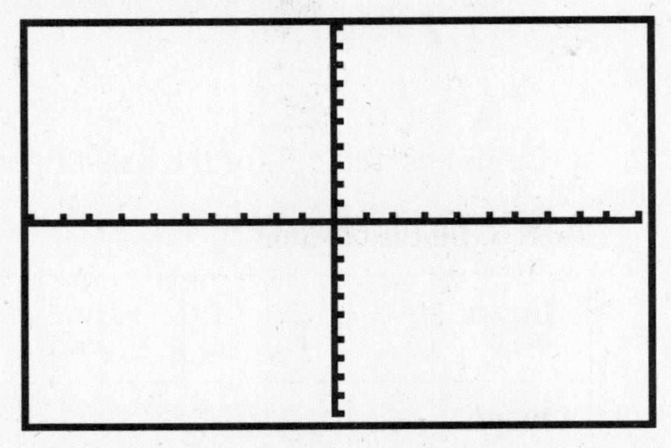

d. Compare and contrast the information you get from the table versus the equation versus the graph.

4. a. Complete Table 5 for the basic oppositing function $Opp(x) = -x$.

Table 5: Oppositing function

Input	–3	–2	–1	$-\frac{1}{2}$	$-\frac{1}{4}$	0	$\frac{1}{4}$	$\frac{1}{2}$	1	2	3
Output											

b. Write the ordered pairs from Table 5 where the input is not an integer.

c. Sketch the graph of the function. Clearly label your graph with the ordered pairs from the table.

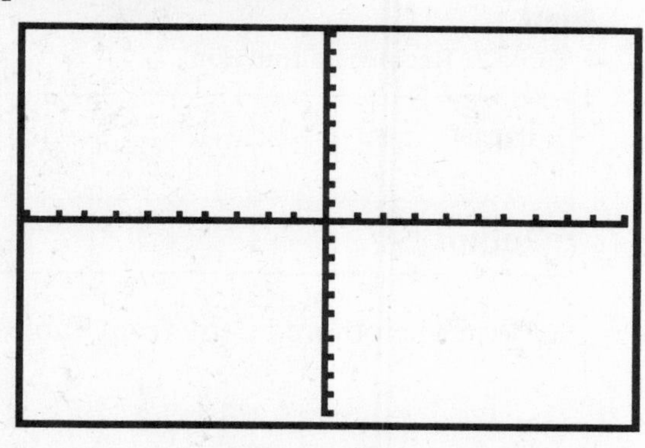

d. Compare and contrast the information you get from the table versus the equation versus the graph.

5. a. Complete Table 6 for the basic absolute value function $Abs(x) = |x|$.

Table 6: Absolute value function

Input	−3	−2	−1	$-\frac{1}{2}$	$-\frac{1}{4}$	0	$\frac{1}{4}$	$\frac{1}{2}$	1	2	3
Output											

b. Write the ordered pairs from Table 6 where the output is larger than 2.

c. Sketch the graph of the function. Clearly label your graph with the ordered pairs from the table.

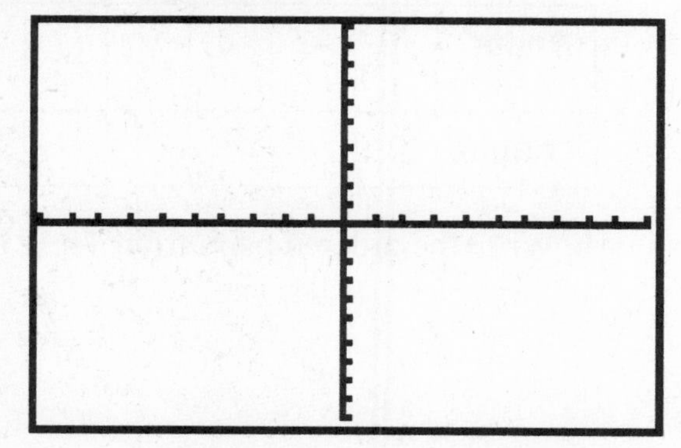

6. a. Complete Table 7 for the basic reciprocal function $Rec(x) = \frac{1}{x}$.

Table 7: Reciprocal function

Input	–3	–2	–1	$-\frac{1}{2}$	$-\frac{1}{4}$	0	$\frac{1}{4}$	$\frac{1}{2}$	1	2	3
Output											

b. Write the ordered pairs from Table 7 where the output is an integer.

c. Sketch the graph of the function. Clearly label your graph with the ordered pairs from the table.

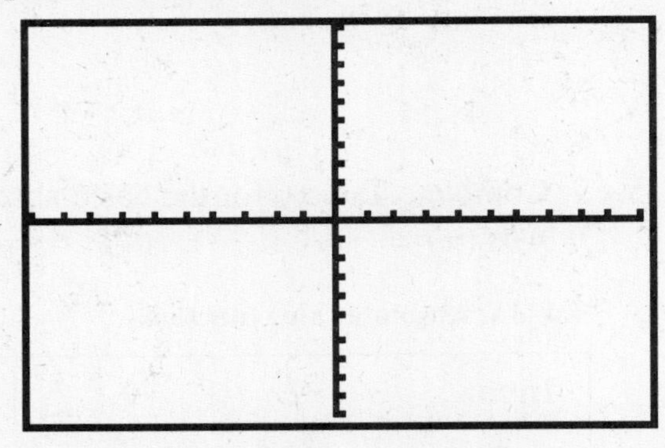

d. Compare and contrast the information you get from the table versus the equation versus the graph.

7. Complete Table 8 for the basic square root function $Sqrt(x) = \sqrt{x}$.

Table 8: Square root function

Input	–3	–2	–1	$-\frac{1}{2}$	$-\frac{1}{4}$	0	$\frac{1}{4}$	$\frac{1}{2}$	1	2	3
Output											

b. Write the ordered pairs from Table 8 where the output is less than 1.

c. Sketch the graph of the function. Clearly label your graph with all ordered pairs from the table.

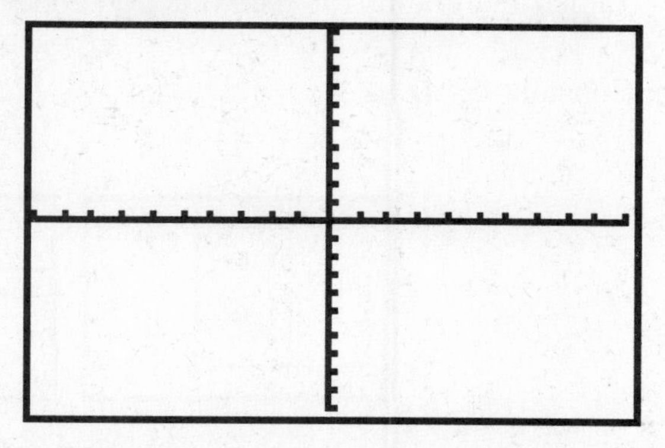

d. Compare and contrast the information you get from the table versus the equation versus the graph.

8. For each of the basic functions, write the domain; that is, list all real numbers that can be used as input.

   a. Constant:

   b. Linear:

   c. Quadratic:

   d. Oppositing:

   e. Absolute value:

   f. Reciprocal:

   g. Square root:

9. For each of the basic functions, write the range; that is, list all real numbers that will be output by the function.

   a. Constant:

   b. Linear:

   c. Quadratic:

   d. Oppositing:

   e. Absolute value:

   f. Reciprocal:

   g. Square root:

Discussion

Tables and graphs for each function appear in Figures 1–7:

Constant $C(x) = 3$:

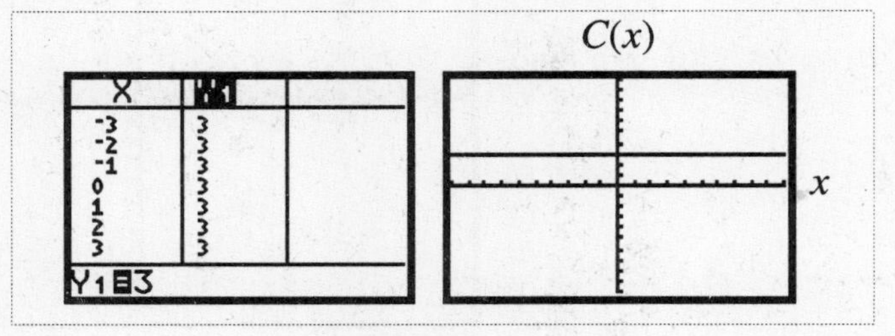

Figure 1

Linear $L(x) = x$:

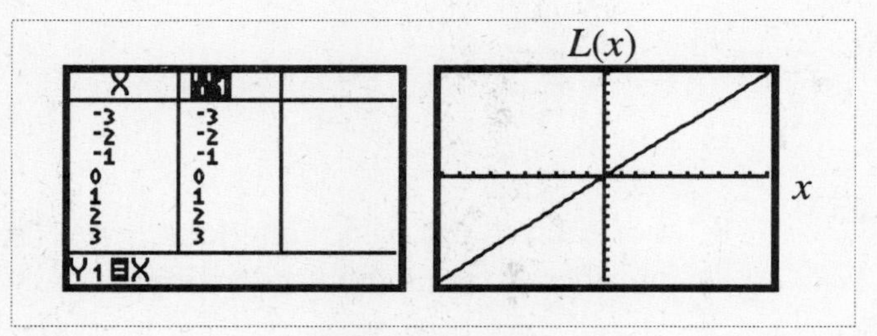

Figure 2

Quadratic $Q(x) = x^2$:

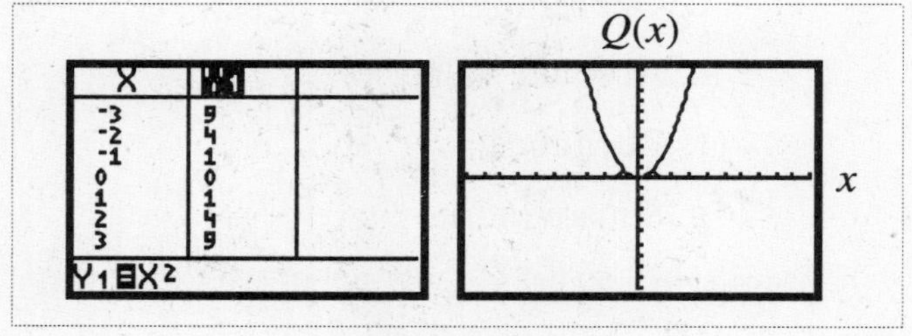

Figure 3

Oppositing Function $Opp(x) = -x$:

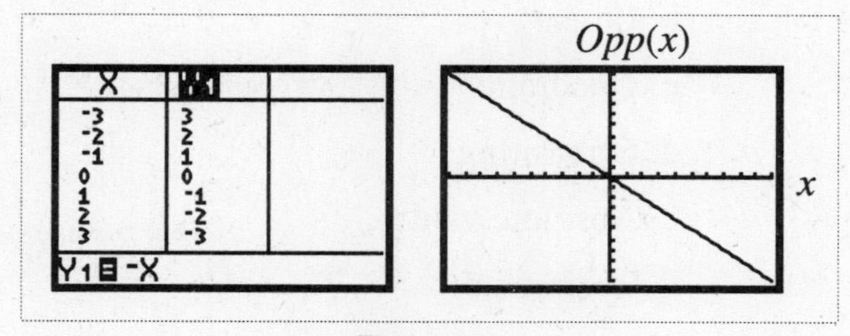

Figure 4

Absolute Value Function $Abs(x) = |x|$:

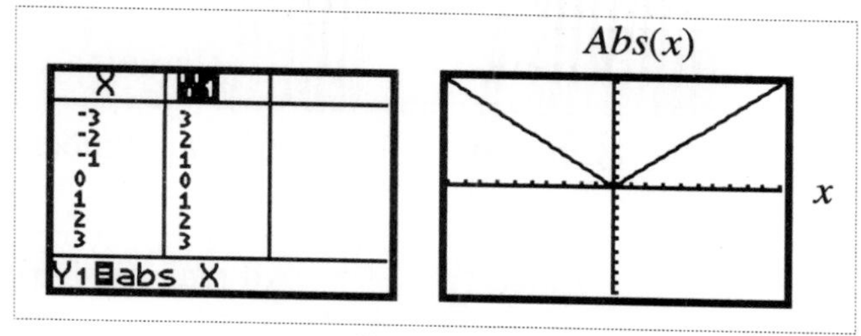

Figure 5

Reciprocal Function $Rec(x) = \frac{1}{x}$:

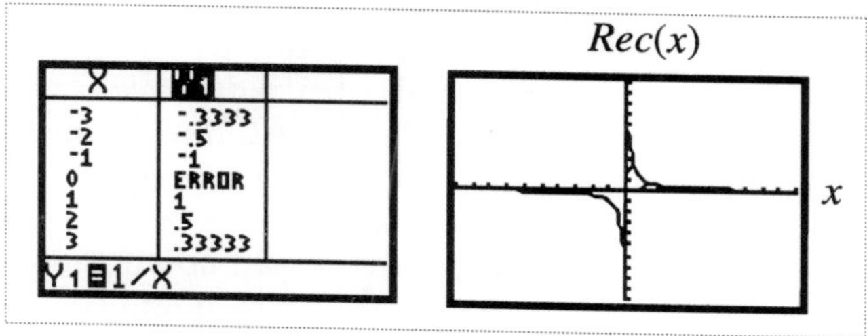

Figure 6

Square Root Function $Sqrt(x) = \sqrt{x}$:

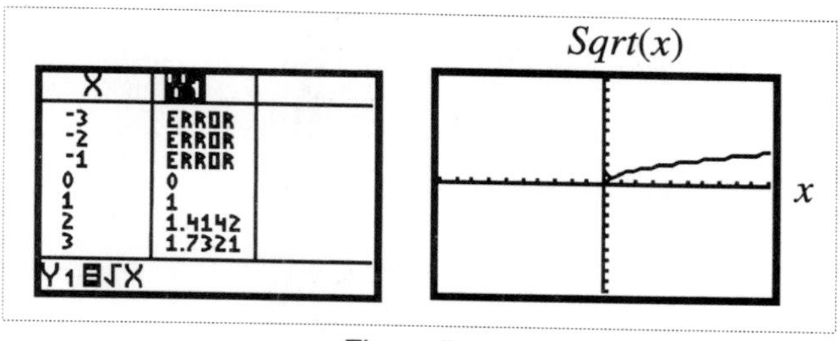

Figure 7

The set of all inputs to a function is called the ***domain of the function***. Using a graph we can estimate the domain by noting all values on the horizontal axis that have output values.

The set of all outputs of a function is called the ***range of the function***. Using a graph the range can be estimated by noting all values on the vertical axis that have corresponding input values.

Table 9 identifies the domain and range of the basic functions.

Function	Domain	Range
Constant	All real numbers	The value of the constant
Linear	All real numbers	All real numbers
Quadratic	All real numbers	All nonnegative real numbers
Oppositing	All real numbers	All real numbers
Absolute value	All real numbers	All nonnegative real numbers
Reciprocal	All real numbers except 0	All real numbers except 0
Square root	All nonnegative real numbers	All nonnegative real numbers

Table 9: Domain and range of basic functions

Investigation

10. For each basic function, use Tables 2–8 to identify the input value where the output is zero.

    a. Constant:

    b. Linear:

    c. Quadratic:

    d. Oppositing:

    e. Absolute value:

    f. Reciprocal:

    g. Square root:

11. Graph each function in a "friendly" viewing window. Find the point on the graph where the output is zero. Write the ordered pair for the point.

    a. Constant:

    b. Linear:

    c. Quadratic:

    d. Oppositing:

e. Absolute value:

f. Reciprocal:

g. Square root:

12. How do your answers to Investigation 11 compare to the answers to Investigation 10?

13. For each basic function, use Tables 2–8 to identify the output value where the input is zero.

    a. Constant:

    b. Linear:

    c. Quadratic:

    d. Oppositing:

    e. Absolute value:

    f. Reciprocal:

    g. Square root:

14. Graph each function in a "friendly" viewing window. Find the point on the graph where the input is zero. Write the ordered pair for the point.

    a. Constant:

    b. Linear:

    c. Quadratic:

    d. Oppositing:

    e. Absolute value:

    f. Reciprocal:

    g. Square root:

15. How do your answers to Investigation 14 compare to the answers to Investigation 13?

16. Consider the function $y(x) = 4x - 8$.

    a. Find the output if the input is zero.

    b. Use a table to find the input where the output is zero.

    c. Graph the function in a "friendly" viewing window. Find the point where the output is zero. Express the point as an ordered pair. Describe where you looked on the graph to answer this question.

17. Consider the function $y(x) = x^2 - 6x + 8$.

    a. Find the output if the input is zero.

    b. Use a table to find the inputs where the output is zero.

    c. Graph the function in a "friendly" viewing window. Find the points where the output is zero. Express the points as ordered pairs. Describe where you looked on the graph to answer this question.

Discussion

The output where the **input** is **zero** is called the ***vertical intercept*** (sometimes known as the ***y–intercept***). This corresponds to the point where the graph of the function intersects the output axis.

The input that causes the **output** to be **zero** is called the ***horizontal intercept*** (sometimes known as the ***x–intercept***). This corresponds to the point where the graph of the function intersects the input axis. Horizontal intercepts are also called the ***zeros of the function***.

Table 10 displays the vertical intercepts and horizontal intercepts for the basic functions.

Function	Vertical intercept	Horizontal intercept
Constant	3	None
Linear	0	0
Quadratic	0	0
Oppositing	0	0
Absolute value	0	0
Reciprocal	None	None
Square root	0	0

Table 10: Intercepts

Other linear and quadratic functions are built from the basic constant, linear and quadratic functions using addition, subtraction, and multiplication.

The most ***general linear function*** is $y(x) = ax + b$ where a and b are real numbers. This function is built by first multiplying the constant function $C(x) = a$ and the identity function $L(x)$ and then adding the constant function $C(x) = b$. The constant function is a special case of the linear function in which $a = 0$. The function $y(x) = 4x - 8$ is an example of a general linear function.

The table and graph of $y(x) = 4x - 8$ appear in Figure 8.

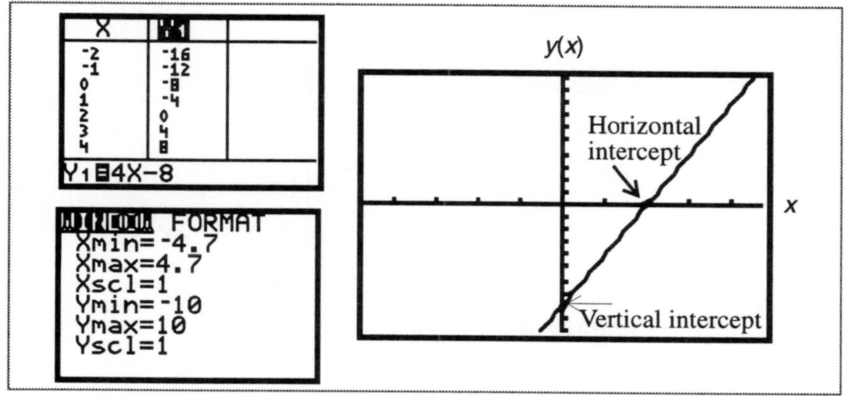

Figure 8

The graph has a vertical intercept at $y = -8$ and a horizontal intercept at $x = 2$. The ordered pair representations are (0, –8) and (2, 0).

The most ***general quadratic function*** has form $y(x) = ax^2 + bx + c$ where *a*, *b*, and *c* are real numbers. This function is built by multiplying the constant function $C(x) = a$ by the quadratic function $Q(x) = x^2$. We then multiply the constant function $C(x) = b$ by the identity function $L(x) = x$. We complete the function by adding the previous two products and the constant function $C(x) = c$. The function $y(x) = x^2 - 6x + 8$ is an example.

The table and graph of $y(x) = x^2 - 6x + 8$ appears in Figure 9.

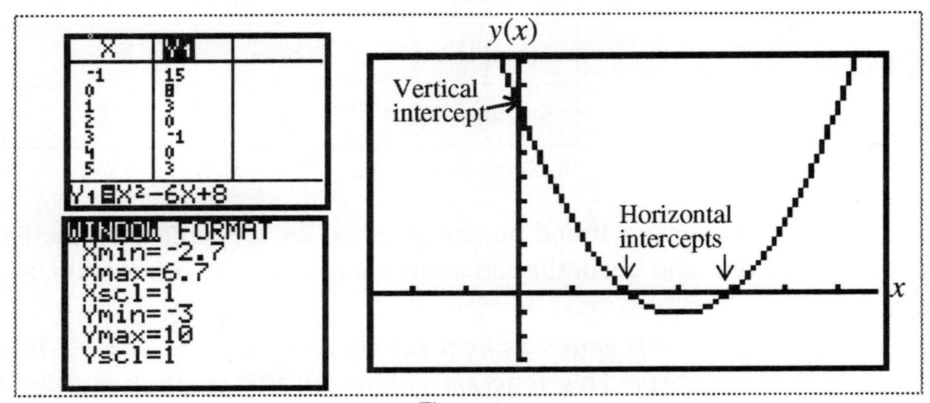

Figure 9

The vertical intercept is at $y = 8$ and the horizontal intercepts are at $x = 2$ and $x = 4$. The ordered pairs are (0, 8), (2, 0), and (4, 0).

Many other classes of functions are built from the basic functions. Table 11 extends Table 1 to include the more general form of the functions discussed in this section. The letters *a*, *b*, and *c* represent real numbers. *f(x)* represents any function of *x*.

Function name	Basic form	General form
Linear (Identity)	$L(x) = x$	$y(x) = ax + b$
Quadratic	$Q(x) = x^2$	$y(x) = ax^2 + bx + c$
Opposite	$Opp(x) = -x$	$y(x) = -f(x)$
Absolute value	$Abs(x) = \|x\|$	$y(x) = \|f(x)\|$
Reciprocal	$Rec(x) = \frac{1}{x}$	$y(x) = \frac{1}{f(x)}$
Square root	$Sqrt(x) = \sqrt{x}$	$y(x) = \sqrt{f(x)}$

Table 11: General form of basic functions

Explorations

1. List and define the words in this section that appear in ***italics bold*** type.

2. For each basic function, let the input increase from 0 to 5.

   a. For which functions does the output increase?

   b. For which functions does the output decrease?

   c. Which function's output increases the fastest?

   d. Which function's output increases the slowest?

   e. Which function's output decreases the fastest?

   f. Which function's output decreases the slowest?

3. Use the graph of the basic function to answer each of the following.

   a. For what value(s) of x is $L(x) = 17$?

   b. For what value(s) of x is $Q(x) = 16$?

   c. For what value(s) of x is $Abs(x) = 13$?

   d. For what value(s) of x is $Opp(x) = 13$?

   e. For what value(s) of x is $Rec(x) = \frac{2}{17}$?

   f. For what value(s) of x is $Sqrt(x) = 3$?

4. For each basic function, identify a real–life story problem in this text that uses this basic function.

5. Consider $y(x) = 2x + 4$.

   a. Complete Table 12:

   Table 12: Input/Output Table

x	–5	–4	–3	–2	–1	0	1	2	3	4	5
$y(x)$											

   b. Use the table to identify the vertical intercept and the horizontal intercept.

   c. Graph the function. What method did you use? Why are you sure your graph is correct?

   d. Is this function increasing or decreasing? Defend your answer.

6. Consider $y(x) = -2x + 6$.

   a. Complete Table 13:

   Table 13: Input/Output Table

x	–5	–4	–3	–2	–1	0	1	2	3	4	5
$y(x)$											

   b. Use the table to identify the vertical intercept and the horizontal intercept.

   c. Graph the function. What method did you use? Why are you sure your graph is correct?

   d. Is this function increasing, decreasing, or neither? Defend your answer.

   e. You are talking to a friend on the phone. Describe the graph to your friend so that she can accurately draw the graph from your description. Be sure to include all important features in your description.

7. Consider $y(x) = x^2 + x - 6$.

   a. Complete Table 14:

   Table 14: Input/Output Table

x	–5	–4	–3	–2	–1	0	1	2	3	4	5
$y(x)$											

   b. Use the table to identify the vertical intercept and the horizontal intercepts.

   c. Graph the function. What method did you use? Why are you sure your graph is correct?

   d. Is this function increasing, decreasing, or neither? Defend your answer.

   e. In Table 14, what are the inputs and what are the outputs?

   f. Is the domain of the function just the set of inputs from Table 14? If not, what is the domain?

   g. Is the range of the function just the set of outputs from Table 14? If not, what is the range?

8. For the functions in Explorations 5–7, how many horizontal intercepts does each have? Why does one of the functions have a different number of horizontal intercepts from the other two?

9. What happens to the value of $\frac{1}{x}$ when x is positive and x grows larger and larger? Justify your answer.

10. A spaceship travels in stages that are the same distance long as shown in Figure 10.

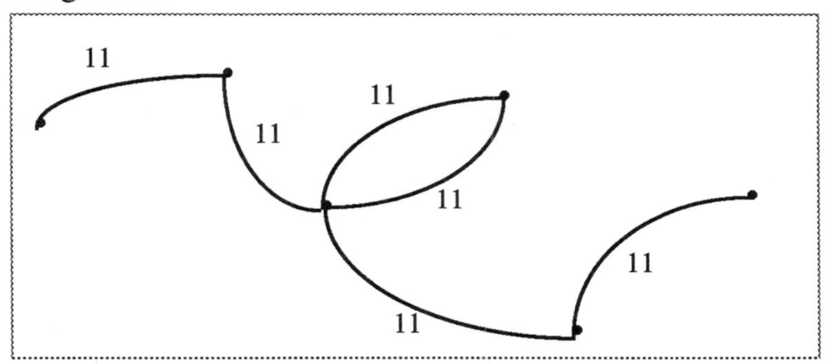

Figure 10

a. If each stage is 11 light–years long, write an expression that represents how far the spaceship will go in y stages.

b. Classify the function that is the answer to part a.

c. Suppose the spaceship begins its travels four light–years from earth. Write a new function that measures the distance of the spaceship from earth after traveling y stages.

Concept Map

Construct a concept map centered on the phrase **basic mathematical functions**.

Reflection

Describe the important ideas about functions that you learned in this section.

Describe the positive and negative aspects of investigating relationships using

a. a table.

b. an equation.

c. a graph.

What do you gain by looking at all three representations? Compare the understanding obtained by looking at only one versus many representations.

Section 6.4 Polynomials

Purpose

- Introduce polynomials as algebraic expressions.
- Investigate the order of operations within a polynomial.
- Investigate operations on polynomials.

Investigation

1. Karen's weekly salary is \$100 plus a 15% commission on her weekly sales.

   a. Describe in words how to compute Karen's weekly salary given her weekly sales.

   b. Draw a function machine with input weekly sales and output weekly salary.

   c. Complete Table 1.

Weekly sales (\$)	Weekly salary (\$)
500	
1000	
1500	
2000	

Table 1: Salary based on sales

   d. If weekly sales x is input and weekly salary S is output, write an algebraic representation (equation) for the relationship $S(x)$.

2. A company sells rectangular sheets of metal. The length of each sheet is three feet more than the width of the sheet.

   a. Describe in words how to find the area of a rectangular sheet given the width.

   b. Construct a function machine with input width and output area.

   c. Complete Table 2.

Width (feet)	**Length (feet)**	**Area (square feet)**
6		
10		
14		
18		

Table 2: Area of rectangles

   d. If x represents the width of a sheet, write an algebraic expression for the length.

   e. The area of each sheet is the product of the length and the width. If the width x is input and the area A is output, write an algebraic representation (equation) for the relationship $A(x)$.

3. a. Evaluate $7y - 5$ when $y = -6$. What is the order of operations? State in words what you did.

b. Evaluate $3x^2 + 5x - 6$ when $x = 2$. What is the order of operations? State in words what you did.

Discussion

The algebraic representations for the problem situations in Investigations 1 and 2 are

$S(x) = 100 + 0.15x$ and $A(x) = x(x+3)$ or $A(x) = x^2 + 3x$.

The function $S(x)$ is a linear function since the highest power on the input variable x is one. The function $A(x)$ is a quadratic function since the highest power on the input variable x is two. Both $100 + 0.15x$ and $x^2 + 3x$ represent real numbers when a value of x is used as input. These two expressions are examples of ***polynomials***.

Well, what is a polynomial?

A ***polynomial*** is a mathematical expression consisting of a sum of ***polynomial terms***.

Great! But what's a polynomial term?

A ***polynomial term*** is a constant or a product of a constant and one or more variables. Note that $7x^3$ is a polynomial term since it can be expressed as $7xxx$, but $7x^{-3}$ is not a polynomial term. $7x^3$ is analogous to the expanded form of the whole number 7000 which is expressed as $7(1000) = 7(10^3)$. For purposes of this section, we will use the word ***term*** in place of polynomial term.

Some polynomials have only one term. In this case the polynomial is called a ***monomial***. The terms -5, $4x^3$, $\frac{2}{3}xyz$, $\sqrt{3}x$, and y are examples of monomials.

The constant factor in each term is called the ***numerical coefficient*** of the term. The constant term is analogous to the units digit in a whole number.

So the terms of $100 + 0.15x$ are 100, a constant, and $0.15x$, the product of 0.15 and x. The numerical coefficient of 100 is 100. The numerical coefficient of $0.15x$ is 0.15.

The terms of $x^2 + 3x$ are x^2, the product of x and x, and $3x$, the product of 3 and x. The numerical coefficient of x^2 is 1 since $x^2 = 1x^2$. The numerical coefficient of $3x$ is 3.

The two expressions in Investigation 3 are polynomials too. Can you identify the terms? Don't just say yes. Do it!

The four polynomials discussed thus far are polynomials in one variable. If the highest power on the variable is one, we call the polynomial a ***linear polynomial***. The one is often implied rather than written. In $100 + 0.15x$, $0.15x = 0.15x^1$. Note that a linear function is a function whose process results in a linear polynomial. If the highest power on the variable is two, we call the polynomial a ***quadratic polynomial***. A quadratic function is a function whose process results in a quadratic polynomial.

$100 + 0.15x$ and $7y - 5$ are examples of linear polynomials.

$x^2 + 3x$ and $3x^2 + 5x - 6$ are examples of quadratic polynomials.

The highest power on the variable is called the ***degree*** of the polynomial.

Linear polynomials have a degree of 1.

Quadratic polynomials have a degree of 2.

Polynomials are important in mathematics since they are used to represent models of real–life problems. Many of the algebraic representations we have previously considered contained polynomials.

Since polynomials represent real numbers, we need to be able to simplify and manipulate them just as we do with numbers. Let's investigate how to do this.

As you do the following operations, compare with how you add, subtract, and multiply whole numbers.

For example,

$$\begin{aligned} 241 + 35 &= (2(100) + 4(10) + 1(1)) + (3(10) + 5(1)) \\ &= (2(10^2) + 4(10^1) + 1(1)) + (3(10^1) + 5(1)) \\ &= 2(10^2) + (4(10^1) + 3(10^1)) + (1(1) + 5(1)) \\ &= 2(10^2) + 7(10^1) + 6(1) \end{aligned}$$

Investigation

4. Predict how each of the following will simplify. Use a symbol manipulator to simplify and compare the answers to your predictions.

Expression	Prediction	Symbol Manipulator
$(3x+2) + (5x+7)$		
$3x + 4xy + 7x - 9xy + 6$		
$(x^2 - 3x + 5) + (7x + 9 - 4x^2)$		
$(3x + 4y - 9) - (5x - 8y + 2)$		
$(3x^2 - 2x - 7) - (x^2 + 3x - 2)$		

Table 3: Adding and subtracting polynomials

5. Study the results of Investigation 4. Simplify the following by hand. Check your answers on a symbol manipulator.

   a. $(2x+3) + (4x-7)$

   b. $(x^2 + 3x - 5) + (4x^2 - x + 2)$

   c. $(5x-9) - (3x-2)$

   d. $(5x^2 - 9x + 3) - (2x^2 + 3x - 4)$

6. What terms can be combined to simplify a polynomial expression?

7. How does the simplification differ when there is a "–" preceding a parenthesis compared to a "+" preceding a parenthesis?

Discussion

In Investigations 4–7, you studied how to add and subtract polynomials. The most important idea is the recognition of ***like terms***.

Like terms are terms that are identical except possibly for their numerical coefficients. They are similar to digits that have the same place value in two whole numbers.

Polynomials are simplified by adding or subtracting the numerical coefficients of like terms. This is called ***combining like terms***. Unlike terms must remain separate.

We are using the distributive property of multiplication over either addition or subtraction when we combine like terms. For example, both terms of the polynomial $3x + 5x$ contain an x. The x has been distributed over the two terms. If we rewrite the polynomial so the x has not been distributed, we get $(3 + 5)x$, which is equal to $8x$.

$$3x + 5x = (3 + 5)x = 8x$$

Another important idea when subtracting is the operation of oppositing. Recall that a subtraction can be rewritten as an addition by changing the operation from subtraction to addition and finding the opposite of the subtrahend. Polynomials are subtracted in a similar manner.

The opposite of a polynomial is found by changing the sign on every term in the polynomial. We are using the polynomial as input to the oppositing function machine.

For example, $-(2x^2 - 5x + 3) = -2x^2 + 5x - 3$. Verify these are equivalent by comparing their outputs in a table or by graphing both functions. Both the table outputs and the graphs are identical!

You can determine if two expressions in one variable are equivalent by comparing their outputs in a table. The outputs are the same if the expressions are equivalent.

Problem Solving Toolkit

You can determine if two expressions in one variable are equivalent by graphing each. The graphs are the same if the expressions are equivalent.

Problem Solving Toolkit

This is an application of the distributive property of multiplication over addition since finding an opposite is equivalent to multiplying by negative one.

Lets look at the problems you did in Investigation 5.

$(2x + 3) + (4x - 7)$	$= 2x + 3 + 4x - 7$	Remove parentheses
	$= 2x + 4x + 3 - 7$	Commute 3 and $4x$
	$= (2 + 4)\,x + 3 - 7$	Distributive property
	$= 6x - 4$	Combine like terms.

$(x^2 + 3x - 5) + (4x^2 - x + 2)$

$= x^2 + 3x - 5 + 4x^2 - x + 2$
Remove parentheses

$= x^2 + 4x^2 + 3x - x - 5 + 2$
Commute terms

$= (1 + 4)\,x^2 + (3 - 1)\,x - 5 + 2$
Distributive property

$= 5x^2 + 2x - 3$ Combine like terms

$$(2x+3)\ (x^2+5x-4)\ =$$

$$(2x)\ (x^2) + (2x)\ (5x) - (2x)\ (4) + (3)\ (x^2) + (3)\ (5x) - (3)\ (4)$$
Distribute

$$= 2x^3 + 10x^2 - 8x + 3x^2 + 15x - 12$$
Multiply like factors

$$= 2x^3 + 13x^2 + 7x - 12$$ Add like terms

Don't forget that tables are an excellent way to check answers. Figure 4 displays a check by table of the last problem.

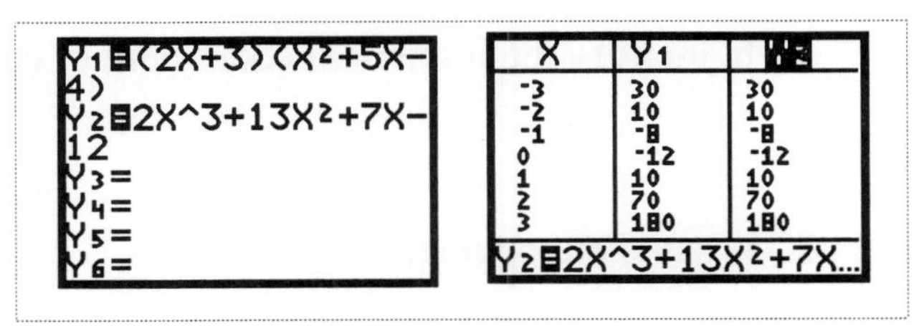

Figure 4

Well, that was fun! Now you get to do some. Make polynomials your friends. They are one of the primary building blocks in applying algebra to life around you.

Explorations

1. List and define the words in this section that appear in ***italics bold*** type.

2. Perform the following operations and simplify completely.

   a. $(x^2-3x+2)+(2x^2-x-7)$ b. $(3x-8)-(2x+3)$

   c. $(x^2+9x-3)-(4x^2+3x+2)$ d. $3x^2(2x-5)$

   e. $-5x(2x^2-7x+9)$ f. $(x+4)\ (x-9)$

   g. $(3x-2)\ (4x+7)$ h. $(7x+2)\ (3x+4)$

   i. $(3x+5)^2$ j. $(x-7)\ (x+7)$

   k. $(4x+3)\ (4x-3)$ l. $(2x-11)^2$

   m. $(2x+3)\ (x^3-7x^2-4x+5)$ n. $(x-1)\ (x^3+x^2+x+1)$

3. Write an example of a binomial.

4. Write an example of a trinomial.

5. Given the linear polynomial $-9x+7$,

   a. identify each term.

   b. identify the numerical coefficient of each term.

6. Give the quadratic polynomial $4x^2 - x + 3$,

   a. identify each term.

   b. identify the numerical coefficient of each term.

7. What is the opposite of $4x^3 - 7x^2 - 9x + 5$?

8. Write two binomials whose product is the difference of two squares.

9. Write a pair of binomials, if possible, so that the product of the two binomials is a

   a. monomial.

   b. binomial.

   c. trinomial.

   d. quadranomial (a polynomial with four terms).

   e. polynomial with five terms.

10. Write an example of a linear binomial with one variable. Square your answer.

11. You belong to a warehouse club. You receive a 20% discount on your total purchases and pay a 5% membership fee on your purchases.

    a. Is it better to receive the discount before paying the fee or vice versa? Does it matter?

    b. Defend your answer by providing a numerical example.

    c. Prove your answer in general using polynomials.

12. The following polynomials resulted from multiplying two binomials. Find the binomials.

    a. $x^2 - 4$

    b. $x^2 + 2x + 1$

    c. $x^2 + 5x + 6$

    d. $x^2 - 12x + 35$

    e. $9x^2 - 25$

    f. $x^2 + 12x + 36$

    g. $x^2 + 2x - 15$

13. Find the output of the linear function $y(r) = 11r - 4$ when $r = -3$.

14. Find the output of the quadratic function $y(x) = x^2 - 7x - 4$ when $x = \frac{2}{3}$.

15. Sandy has a monthly salary of $400 plus a 12% commission on her monthly sales.

    a. Find Sandy's monthly income if her monthly sales are $2763. Which representation (table, equation, or graph) did you use to answer this question? How would you use the other two representations to answer this question?

    b. Find Sandy's monthly sales if she wants to earn $775 this month. Which representation (table, equation, or graph) did you use to answer this question? How would you use the other two representations to answer this question?

16. A company sells rectangular sheets of cardboard. The length of each sheet is seven inches less than the width.

    a. Complete Table 10.

Width (inches)	**Length (inches)**	**Area (square inches)**
30		
35		
50		
62		
w		

Table 10: Area of rectangular sheets of cardboard

b. If the width w represents the input and the area A represents the output, write an algebraic representation (equation) for $A(w)$.

c. Use your answer to part b to find the area when the width is 75 inches.

d. Graph the function from part b on a graphing utility. Use TRACE to find width of a rectangle if the area is 1,550 square inches. Verify your answer with a table.

17. The graph in Figure 5 demonstrates the relationship between the original price (input) and the sale price (output) of goods that are marked down 37%.

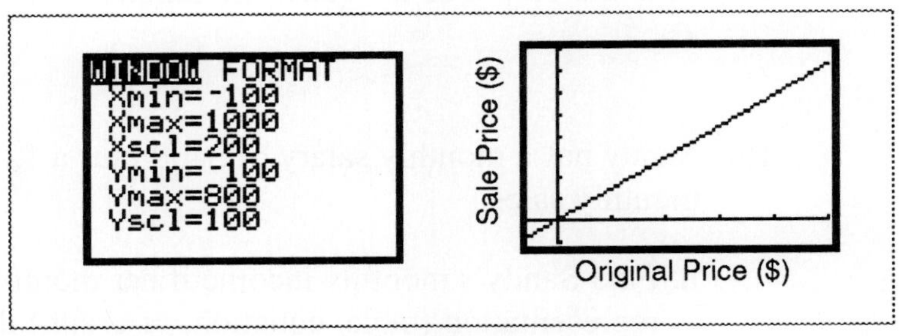

Figure 5

a. What can you tell about the relationship between original price and sale price by looking at the graph?

b. Create a table and record five different input/output pairs.

c. What is the slope of the line in Figure 5?

d. If x dollars represents the input and y dollars represents the output, write an algebraic representation (equation) for $y(x)$.

18. A piece of string is x feet long. It is bent to form a square.

a. Express the length of each side of the square in terms of x. Is this a linear or quadratic function?

b. Express the area of the square in terms of x. Is this a linear or quadratic function?

19. I. M. Rich invested $10,000, some at an annual interest rate of 8% and the rest at an annual interest rate of 6%.

a. If I. M. invested $4,000 at 8%, what would her interest earnings on the $4000 be at the end of one year?

b. If I. M. invested $4,000 at 8%, how much did she invest at 6%?

c. If I. M. invested \$4,000 at 8%, what would her interest earnings on the amount invested at 6% be at the end of one year?

d. If I. M. invested \$4,000 at 8%, what would her total interest earnings be at the end of one year?

e. Complete Table 11. All amounts are in dollars.

Amount invested at 8%	Interest earned on 8% investment after one year	Amount invested at 6%	Interest earned on 6% investment after one year	Total interest earned after one year
0				
3000				
6000				
9000				
x				

Table 11: I. M. Rich's investments

f. If x dollars invested at 8% is the input and total interest T is the output, write an algebraic representation (equation) for $T(x)$.

g. What is the slope of the graph of the function defined in part f?

h. On a graphing utility, graph the function from part f. Determine how much must be invested at 8% to realize interest earnings of \$760. Verify your graphical answer with a table.

Concept Map

Construct a concept map centered on the word **polynomial**.

Reflection

Explain in your own words how to add two polynomials.

Explain in your own words how to multiply two polynomials.

Write an example of a mathematical expression that is not a polynomial. Justify.

Describe how to differentiate between a term and a factor.

Section 6.5 Making Connections: Is there *REAL*ly a completion to the number system?

Purpose

- Reflect upon ideas explored in Chapter 6.
- Explore the connections between a function and its domain.

Investigation

In this section you will work outside the system to reflect upon the mathematics in Chapter 6: what you've done and how you've done it.

1. State the five most important ideas in this chapter. Why did you select each?

2. Identify all the mathematical concepts, processes and skills you used to investigate the problems in Chapter 6?

3. What has been common to all of the investigations which you have completed?

4. Select a key idea from this chapter. Write a paragraph explaining it to a confused best friend.

5. You have investigated many problems in this chapter.

   a. List your three favorite problems and tell why you selected each of them.

   b. Which problem did you think was the most difficult when you began working on it?

   c. What helped you make sense of this problem?

Discussion

There are a number of really important ideas which you might have listed including real number, polynomials, square roots, exponents, linear functions, quadratic function and other basic functions.

Concept Map

Construct a concept map centered around the word **domain.**

Reflection

How does the study of a basic function help you analyze a more general function of the same degree?

Illustration

Draw a picture of **a mathematics test-taker** (a person taking a mathematics test).

The
Test - Taker

Chapter 7

Answering Questions with Linear and Quadratic Functions

Section 7.1 Linear Functions

Purpose

- Illustrate important relationships between numeric, algebraic, and graphical representations of linear functions.
- Investigate the slope of linear functions.
- Investigate the intercepts of linear functions.

Investigation

Consider the problem: Two business majors, Sue and Tom, sell cookies during breaks between classes. They spend $30 on supplies and sell cookies for 25¢ each.

We wish to investigate the relationship between the number of cookies sold (input) and the net profit (output).

1. Describe in words how to calculate the profit given the number of cookies sold.

2. Construct a function machine with input the number of cookies sold and output profit.

3. Complete Table 1.

Number of cookies sold	Net profit (dollars)
0	
50	
100	
150	
200	
c	

Table 1: Number of cookies sold versus net profit

4. Let c represent the number of cookies sold and P represent the net profit, write an algebraic representation (equation) for $P(c)$.

5. What is the finite difference Δc of the inputs in Table 1? Show a computation to justify your answer. What does Δc represent?

6. What is the finite difference ΔP of the outputs in Table 1? Show a computation to justify your answer. What does ΔP represent?

7. What is the ratio of ΔP to Δc? Describe the meaning of this ratio in words.

8. Graph the function $P(c)$. What is the vertical intercept of the graph?

Explorations

1. List and define the words in this section that appear in ***italics bold*** type

For each linear function in Explorations 2–12, identify the

a. slope.

b. output at the vertical intercept.

c. input at the horizontal intercept.

d. graph as increasing or decreasing.

2. $y(x) = 3x - 12$

3. $y(x) = -5x + 10$

4. $y(x) = -3$

5. $y(t) = 7t + 21$

6. $k(r) = -2r - 5$

7. $p(r) = \frac{3}{7}r + 2$

8. $y(t) = 0$

9. $l(t) = \frac{2}{3}t - 5$

10. $m(p) = 7p + \frac{8}{5}$

11. $y(x) = -\frac{3}{4}x + \frac{2}{3}$

12. $z(s) = \frac{5}{7}s - 2\frac{3}{4}$

Using the information in Explorations 13–17,

a. graph the line.

b. write the equation of the line.

c. identify the input at the horizontal intercept.

13. Slope = 3; Vertical intercept = (0, 4).

14. Slope = –2; Vertical intercept = (0, 3).

15. Slope = $\frac{3}{2}$; Vertical intercept = (0, –4).

16. Slope = $-\frac{3}{5}$; Vertical intercept = (0, –1).

17. Slope = $\frac{7}{4}$; Vertical intercept = $(0, -\frac{3}{2})$.

18. Assume that the input variable is x and the output variable is y.

    a. Write the equation of the input axis.

    b. Write the equation of the output axis.

    c. Why do these answers make sense?

19. Find the slope of the line containing the two given points.

    a. (3, 7) and (2, 4)

    b. (–2, 6) and (5, –3)

    c. (4, 2) and (9, 6)

20. For the cookie problem (page 339) discussed earlier, what is the input at the horizontal intercept? Why is this value important to the two business majors?

Given the graphs in Explorations 21–23,

a. identify the vertical intercept.

b. identify the slope.

c. write the equation of the line.

d. identify the horizontal intercept.

21.

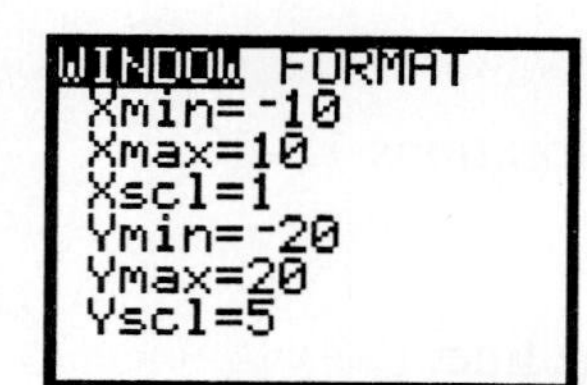

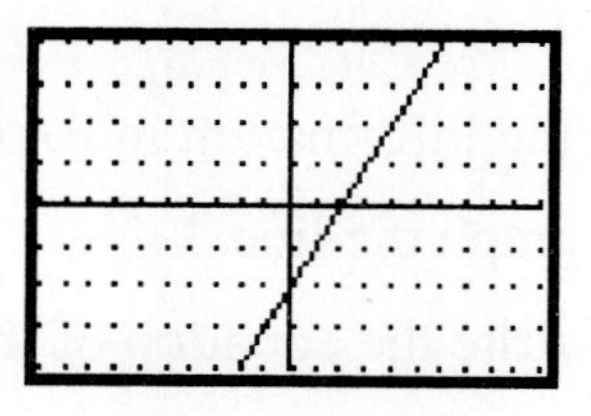

22.

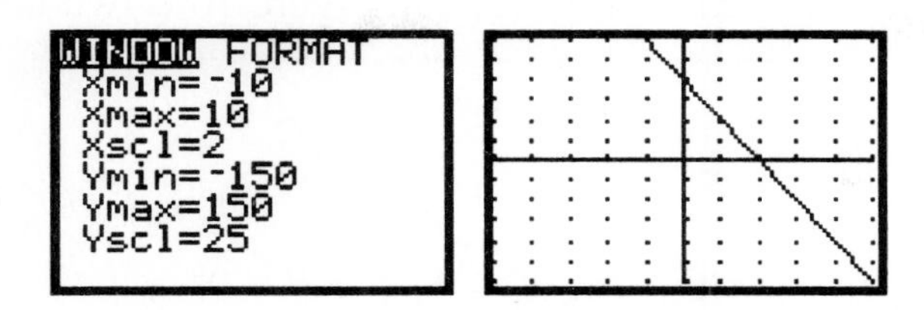

23.

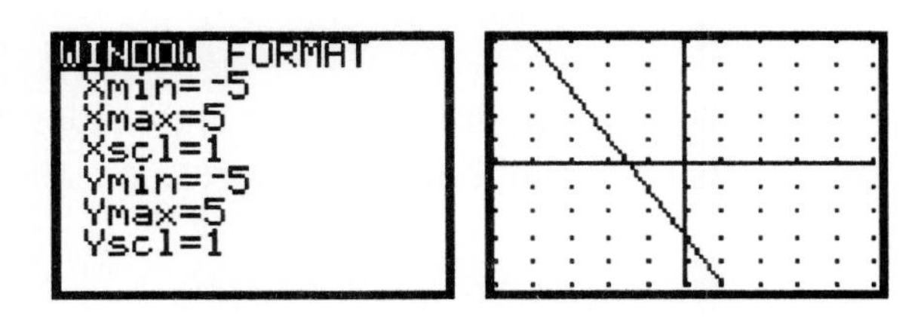

24. To rent a car for a day, WeGotcha Rental charges \$15 plus 20¢ per mile.

   a. If m represents the mileage and T represents the total charge, write an algebraic representation (equation) for $T(m)$.

   b. What is the output at the vertical intercept of this function? What does it represent in the car rental situation?

   c. What is the slope of this function? What does it represent in the car rental situation?

25. Given $y(x) = 2x - 5$,

   a. for what value of x will $y(x) = 0$?

   b. for what value of x will $y(x) = 3$?

   c. for what value of x will $y(x) = 13$?

26. Refer to the function that represents the cookie problem (page 341).

   a. How many cookies must be sold to realize a profit of \$10? Explain your method.

   b. How many cookies must be sold to realize a profit of \$100? Explain your method.

27. Each segment of the shape in Figure 10 is two units in length.

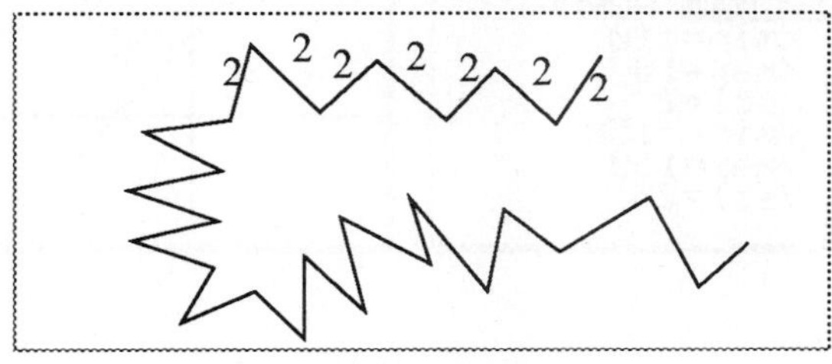

Figure 10

Part of the shape is not drawn. There are n sides altogether each of length 2. Write a mathematical expression for the perimeter of this shape.

28. Consider a linear non–constant function.

a. Can any whole number be used as input? as output?

b. Can any integer be used as input? as output?

c. Can any rational number be used as input? as output?

d. Can any irrational number be used as input? as output?

e. What is the domain of any linear (non–constant) function?

f. What is the range of any linear (non–constant) function?

Concept Map

Construct a concept map centered on the phrase **linear function**.

Reflection

Use your concept map to help you describe everything you know about the general linear function $y(x) = mx + b$.

Section 7.2 Quadratic Functions

Purpose

- Define the general quadratic function.
- Illustrate important relationships between quadratic functions and parabolas in the plane.
- Investigate the intercepts and the vertex of a quadratic function.
- Establish a relationship between the zeros of a quadratic function and the factors of a quadratic function.

Investigation

Recall the square dog pen we constructed in Section 6.1. We began with an area of 100 square feet and doubled the area to obtain an area of 200 square feet. Now we will look at some functions based on the area of a square.

1. If x represents the length of the side of the square, write an expression for the area of the square.

2. Using your answer to Investigation 1, write an expression to represent
   a. twice the area.

   b. triple the area.

   c. quadruple the area.

   d. one–half the area.

   e. seven–tenths of the area.

   f. three–tenths of the area.

3. Using your answer to Investigation 1, write an expression to represent

   a. increasing the area by two square feet.

   b. increasing the area by four square feet.

   c. decreasing the area by two square feet.

   d. decreasing the area by six square feet.

4. Using your answer to Investigation 1, write an expression to represent

   a. three more than twice the area.

   b. five less than triple the area.

5. Using your answer to Investigation 1, write an expression to represent

   a. changing the area by a factor of a where a is positive.

   b. c more than the area where c is positive.

6. Interpret the expression $ax^2 + c$ in terms of the original area of the dog pen. Assume that a and c are positive numbers.

Discussion

Table 1 displays the expressions for the previous investigations if x^2 represents the area.

English statement	Expression
Area	x^2
Twice the area	$2x^2$
Triple the area	$3x^2$
Quadruple the area	$4x^2$
One–half the area	$0.5x^2$
Seven–tenths of the area	$0.7x^2$
Three–tenths of the area	$0.3x^2$
Increasing the area by two square feet	$x^2 + 2$
Increasing the area by four square feet	$x^2 + 4$
Decreasing the area by two square feet	$x^2 - 2$
Decreasing the area by six square units	$x^2 - 6$
Three more than twice the area	$2x^2 + 3$
Five less than triple the area	$3x^2 - 5$

Table 1: Modifying the area of the dog pen

Each of these expressions has a general form $ax^2 + c$. These are examples of quadratic functions. In the following investigations, we look at the graph of $y(x) = ax^2 + c$ by comparing with the graph of the basic quadratic function $Q(x) = x^2$.

Investigation

7. Use a graphing utility to graph $Q(x) = x^2$, $y(x) = 2x^2$, $y(x) = 3x^2$, and $y(x) = 4x^2$ in a "friendly" viewing window. Sketch and label the graphs.

8. a. For each of the functions in Investigation 7, find the output when the input is –1. Find the output when the input is 1.

   b. Describe how the graphs of $y(x) = 2x^2$, $y(x) = 3x^2$, and $y(x) = 4x^2$ compare to the graph of the basic quadratic function $Q(x) = x^2$.

9. Use a graphing utility to graph $Q(x) = x^2, y(x) = 0.7x^2$, $y(x) = 0.5x^2$, and $y(x) = 0.3x^2$ in a "friendly" viewing window. Sketch and label the graphs.

10. a. For each of the functions in Investigation 9, find the output when the input is –1. Find the output when the input is 1.

    b. Describe how the graphs of $y(x) = 0.7x^2$, $y(x) = 0.5x^2$, and $y(x) = 0.3x^2$ compare to the graph of the basic quadratic function $Q(x) = x^2$.

11. Using a graphing utility, graph on the same rectangular grid $Q(x) = x^2, y(x) = -2x^2, y(x) = -3x^2$, and $y(x) = -4x^2$. Sketch and label the graphs.

12. a. For each of the functions in Investigation 11, find the output when the input is –1. Find the output when the input is 1.

b. Describe how the graphs of $y(x) = -2x^2$, $y(x) = -3x^2$, and $y(x) = -4x^2$ compare to the graph of the basic quadratic function $Q(x) = x^2$.

13. Describe, in general, the effect of a on the graph of $y(x) = ax^2$.

14. Using a graphing utility, graph on the same rectangular grid $Q(x) = x^2, y(x) = x^2 + 2, y(x) = x^2 + 4$, and $y(x) = x^2 + 6$. Sketch and label the graphs.

15. Describe how the graphs of $y(x) = x^2 + 2, y(x) = x^2 + 4$, and $y(x) = x^2 + 6$ compare to the graph of the basic quadratic function $Q(x) = x^2$.

16. Using a graphing utility, graph on the same rectangular grid $Q(x) = x^2, y(x) = x^2 - 2, y(x) = x^2 - 4$, and $y(x) = x^2 - 6$. Sketch and label the graphs.

17. Describe how the graphs of $y(x) = x^2 - 2$, $y(x) = x^2 - 4$, and $y(x) = x^2 - 6$ compare to the graph of the basic quadratic function $Q(x) = x^2$.

18. Describe the effect of c on the graph of $y(x) = x^2 + c$.

Discussion

The graph of $y(x) = ax^2 + c$, where a is not zero, is called a ***parabola***. A parabola looks like one of the two sketches in Figure 1.

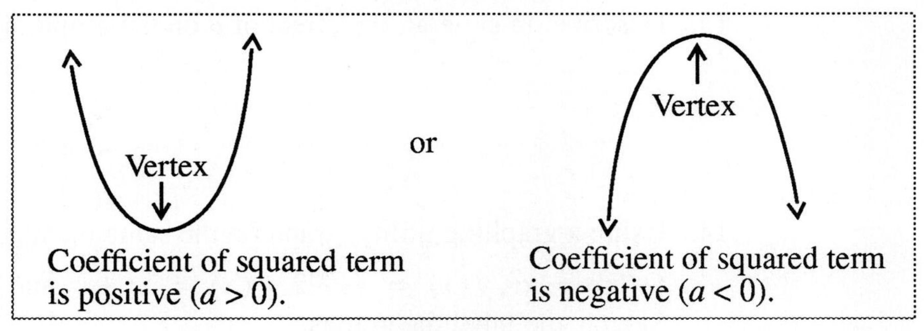

Figure 1

The lowest or highest point is called the ***vertex*** of the parabola. The basic quadratic function $Q(x) = x^2$ has a lowest point at the origin (0, 0). The horizontal intercept is (0, 0) and the vertical intercept is (0, 0). If we draw a vertical line through the vertex and fold the parabola along this line, one branch of the parabola will lie on top of the other branch. This illustrates the ***symmetry*** of the parabola with respect to the vertical line through the vertex.

The graph of $y(x) = ax^2$ where a is larger than one is similar, but rises more quickly.

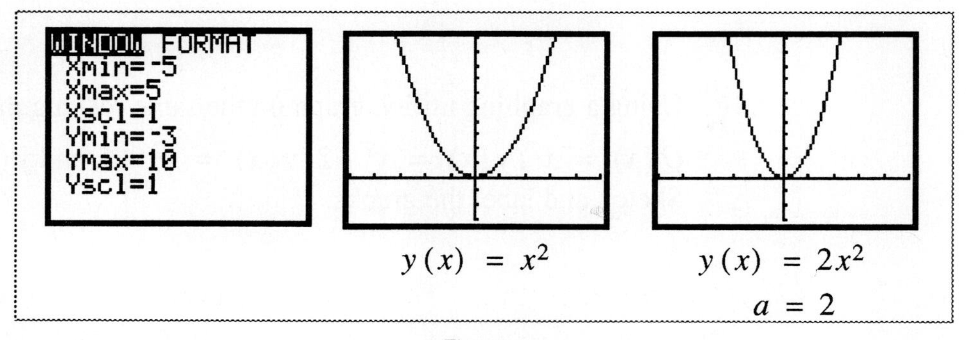

Figure 2

What does it mean to rise more quickly? Let x increase from 0 to 1. The output of $y(x) = x^2$ increases from 0 to 1 while the output of $y(x) = 2x^2$ increases from 0 to 2.

The rate of change of $y(x) = x^2$ as the input increases from 0 to 1 is 1.

$$\frac{\Delta y}{\Delta x} = \frac{1}{1} = 1$$

The rate of change of $y(x) = 2x^2$ as the input increases from 0 to 1 is 2.

$$\frac{\Delta y}{\Delta x} = \frac{2}{1} = 2$$

The graph of $y(x) = ax^2$ where a is between zero and one is similar, but rises more slowly.

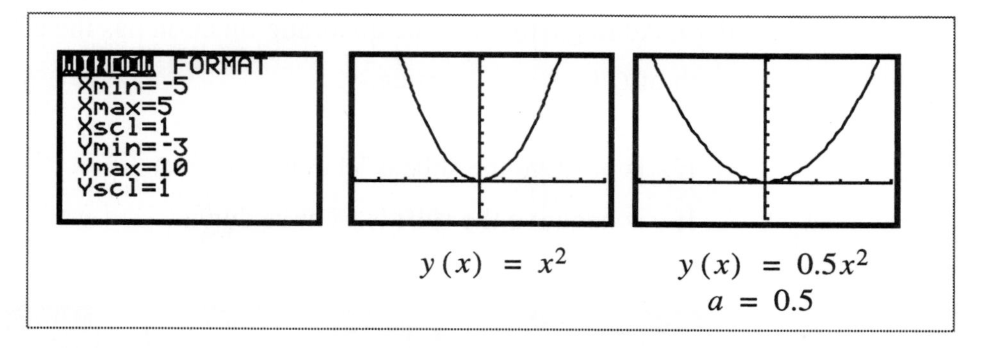

Figure 3

What does it mean to rise more slowly? Let x increase from 0 to 1. The output of $y(x) = x^2$ increases from 0 to 1 while the output of $y(x) = 0.5x^2$ increases from 0 to 0.5.

The rate of change of $y(x) = x^2$ as the input increases from 0 to 1 is 1.

$$\frac{\Delta y}{\Delta x} = \frac{1}{1} = 1$$

The rate of change of $y(x) = 0.5x^2$ as the input increases from 0 to 1 is 0.5.

$$\frac{\Delta y}{\Delta x} = \frac{0.5}{1} = 0.5$$

The graph of $y(x) = ax^2$ where a is negative opens downward, but still has a vertex at (0, 0). Essentially the parabola has been reflected across the horizontal axis.

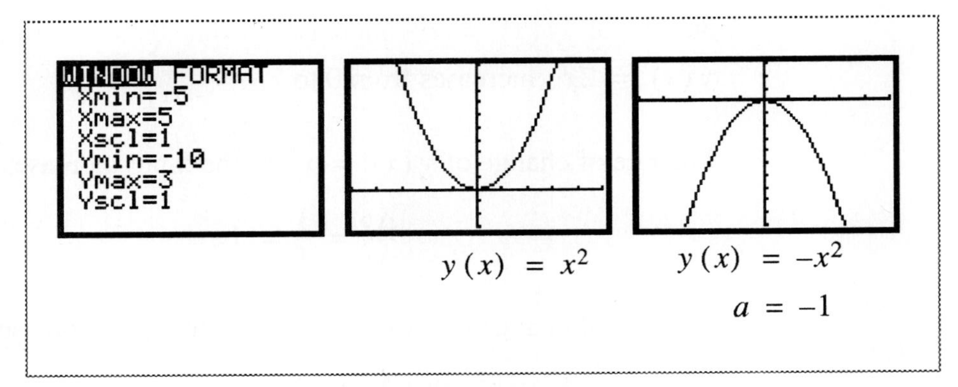

Figure 4

The graph of $Q(x) = x^2$ has a vertical intercept of (0, 0). As you saw in the investigations, the graph of $y(x) = x^2 + c$ has a vertical intercept of (0, c). Adding c to the basic quadratic function has the effect of shifting the graph of $Q(x) = x^2$ vertically.

If c is positive the shift is $|c|$ units up.

If c is negative the shift is $|c|$ units down.

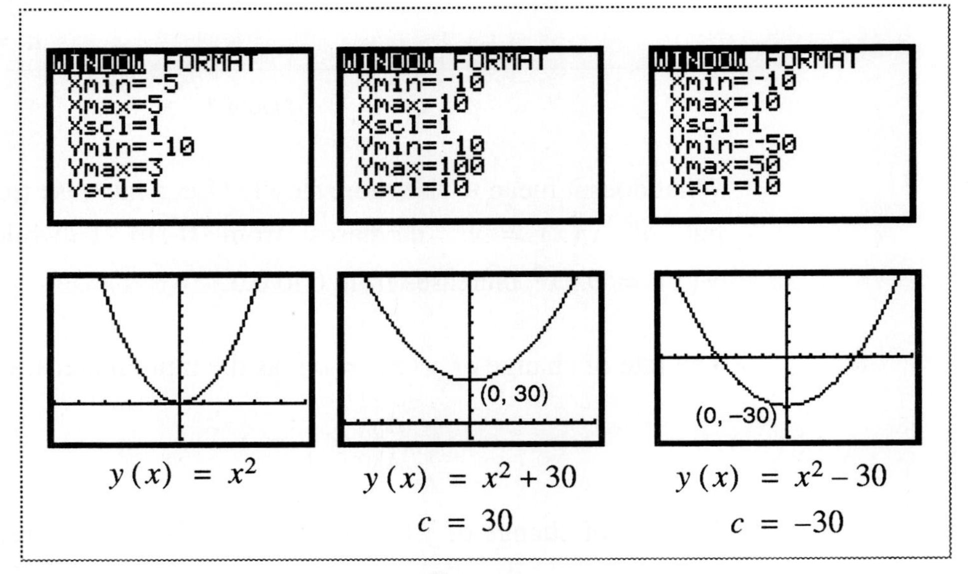

Figure 5

The ***general form of a quadratic function*** is $y(x) = ax^2 + bx + c$ where a, b, and c are real numbers and a is not zero. (Why can't a be zero?) A quadratic function has a degree of two.

Let's investigate this general function further. Note that a, b, and c can be either positive or negative.

Investigation

19. a. Graph $y(x) = x^2 - 4$ on a graphing utility and record the zeros of the function and the x–value of the vertex.

    b. Graph $y(x) = (x+2)(x-2)$ on a graphing utility and record the zeros of this function.

    c. Calculate the output of the functions using the zeros as input.

    d. Display a table for $y(x) = x^2 - 4$ and $y(x) = (x+2)(x-2)$. What do the table outputs tell you about the two functions?

    e. Prove that the two functions are really the same function.

20. a. Graph $y(x) = x^2 + 3x - 4$ on a graphing utility and record the zeros of the function and the x–value of the vertex.

    b. Graph $y(x) = (x+4)(x-1)$ on a graphing utility and record the zeros of this function.

    c. For $y(x) = x^2 + 3x - 4$ and $y(x) = (x+4)(x-1)$, display a table. What do the table outputs tell you about the two functions?

    d. Prove that the two functions are really the same function.

21. a. Graph $y(x) = x^2 - 4x + 4$ on a graphing utility and record the zeros of the function and the x–value of the vertex.

b. Graph $y(x) = (x-2)(x-2)$ on a graphing utility and record the zeros of this function.

c. For $y(x) = x^2 - 4x + 4$ and $y(x) = (x-2)(x-2)$, display a table.What do the table outputs tell you about the two functions?

d. Prove that the two functions are really the same function.

22. Graph $y(x) = x^2 + 4$ on a graphing utility and record the zeros and the x–value of the vertex.

23. In Investigations 19–21, you were given the expanded form and the factored form of a quadratic function. Describe the relationship between the zeros of the function and the factored form of the function. What is different about Investigation 22? Why do you think no factored form was given?

24. How does the location of the x–value of the vertex compare with the location of the zeros? Be specific.

Discussion

In the previous investigations, you saw from the graph that

- $x^2 - 4 = (x-2)(x+2)$, the zeros of this function are 2 and –2, and the x coordinate of the vertex is $x = 0$. We can prove these are the same function by multiplying $(x-2)(x+2)$ out. The result is $x^2 - 4$.

- $x^2 + 3x - 4 = (x+4)(x-1)$, the zeros of this function are –4 and 1, and the x coordinate of the vertex is $x = -\frac{3}{2}$. These functions are equal since the product of $x + 4$ and $x - 1$ is $x^2 + 3x - 4$.

- $x^2 - 4x + 4 = (x-2)(x-2)$, the only zero of this function is 2, and the x coordinate of the vertex is $x = 2$.

- $x^2 + 4$ does not have any real number zeros since the graph does not intersect the horizontal axis. The x coordinate of the vertex is $x = 0$.

The first coordinate of each horizontal intercept of a function, when used as input, produces an output of zero. In addition, the input at the vertex is always the midpoint of the two inputs of the horizontal intercepts.

Let's complete this section by further investigating the relationship between the factored form and the zeros of the function.

Investigation

25. Given the function $y(x) = x^2 + x - 20$,

    a. use a table to identify the zeros.

    b. use the zeros to determine the value of the input at the vertex.

    c. use the zeros to write the function in factored form. Check your answer by multiplying the two factors together.

26. Suppose a quadratic function has 7 and –3 as zeros.

    a. What is the value of the input at the vertex? Why?

    b. Write the factored form of the function. Describe what you did.

    c. Multiply the two factors to obtain the expanded form of the function. Use a table or graph to check if the function has the correct zeros. If it does not, find your mistake.

27. Based on the quadratic functions we have studied,

    a. how many possible zeros will a quadratic function have?

    b. how many possible factors will a quadratic function have?

Discussion

If a quadratic function has a zero at the real number p then one of the ***factors*** of the quadratic function is $(x - (p))$.

The factors of the function have form x subtract the zero p.

The number p has its own sign separate from the subtraction sign following the x in the factor.

For example,

- if 4 is a zero of a quadratic function, then 4 is the value of p and $(x - (4)) = (x - 4)$ is a factor of the quadratic function.
- if –5 is a zero of a quadratic function, then –5 is the value of p and $(x - (-5)) = (x + 5)$ is a factor of the quadratic function.

The function $y(x) = x^2 + x - 20$ has zeros at 4 and –5. Its factored form is $y(x) = (x - 4)(x + 5)$.

If 7 and –3 are zeros of a quadratic function, then the function has factors $(x - (7))$ and $(x - (-3))$. These factors can be written $(x - 7)$ and $(x + 3)$. The quadratic function can be written as $y(x) = (x - 7)(x + 3)$. The expanded form is $y(x) = x^2 - 4x - 21$.

Notice that when we multiply the factors, a linear term may be introduced. This provides a reason for the bx term in the general quadratic function $y(x) = ax^2 + bx + c$.

A quadratic function can have two, one, or no real number zeros. Two zeros arise when the graph has two horizontal intercepts. If the zeros are rational, there will be two distinct factors. One zero occurs when the graph has only one horizontal intercept. In this case, the graph is ***tangent*** to the horizontal axis. If the zero is rational, there are two factors, but they are the same. A quadratic function has no real number zeros if its graph has no horizontal intercepts. We will be unable to factor such a function until we develop a number system beyond the real numbers. If the zeros are irrational, we again will not be able to factor until we study irrational zeros.

When we begin with an expanded form of a quadratic function and obtained the factored form, we are performing a process called ***factoring***. We have seen in this section that a quadratic function can be factored by finding its zeros. The zeros can be used to write the factors. We explore this idea in greater detail in Section 7.5. Time for some explorations.

Explorations

1. List and define the words in this section that appear in ***italics bold*** type.

2. For each of the following quadratic functions, how will the graph compare with the graph of the basic quadratic function $Q(x) = x^2$?

   a. $y(x) = 7x^2$
   b. $y(x) = x^2 + 4$
   c. $y(x) = -5x^2$
   d. $y(x) = x^2 - 7$
   e. $y(x) = 0.4x^2 + 1$
   f. $y(x) = 5x^2 - 4$
   g. $y(x) = -x^2 - 3$

3. Explain why $x - (-6)$ is equivalent to $x + 6$.

4. Given the factored forms of the following quadratic functions, identify the zeros and the x–values of the vertex.

   a. $y(x) = (x-6)(x+6)$

   b. $y(x) = (x-9)(x+7)$

   c. $y(x) = (x+3)(x+5)$

   d. $y(x) = (x-11)(x-7)$

   e. $y(x) = (x+5)(x-8)$

   f. $y(x) = (x-2)(x-2)$

5. Given the following zeros of a quadratic function, write the equation of the function in factored form. Then multiply the factors to obtain the expanded form. Verify your answers using both a table and a graph.

   a. Zeros: 1, 4.

   b. Zeros: –3, –2.

   c. Zeros: 4, –6.

   d. Zeros: –5, 7.

   e. Zeros: –5, 5.

   f. Zero: 7 only.

6. Given the following graphs of quadratic functions, find the zeros, the factored form of the quadratic functions, and the expanded form of the quadratic function.

   a.

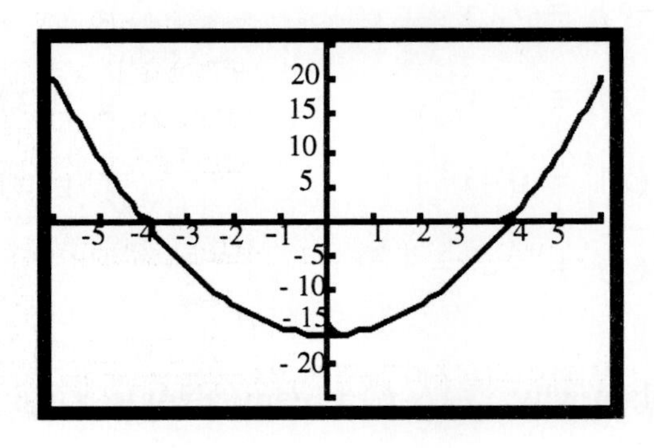

b.

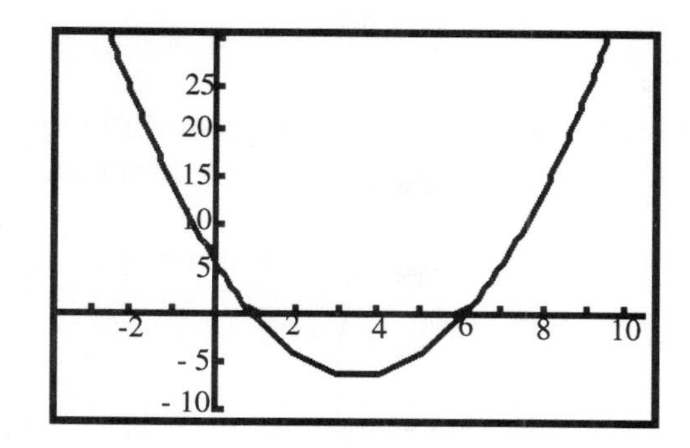

c.

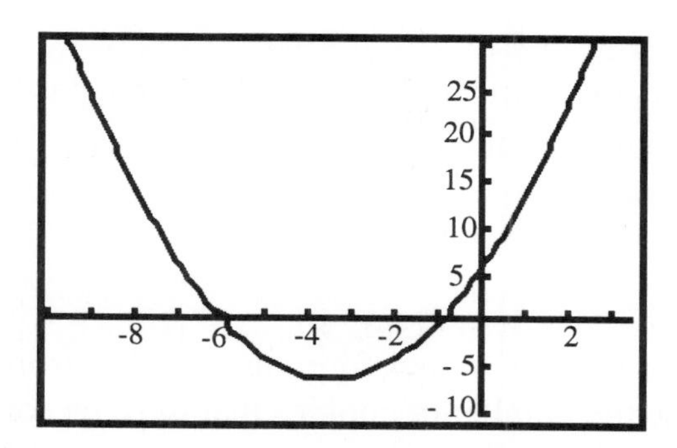

d. You are talking to your friend on the phone again. Describe the graph in part c so that your friend can construct the graph from your description. Be sure to include all important features in your description.

Concept Map

Construct a concept map centered on the phrase **quadratic function**.

Reflection

Describe what is meant by the zeros of a quadratic function.

Describe what is meant by the vertex of a quadratic function.

If $y(x) = (x-s)(x-t)$, what are the zeros of the function?

If the zeros of a quadratic function are a and b, write the function in factored form.

Section 7.3 Linear Equations and Inequalities in One Variable

Purpose

- Investigate the relationship between an equation/inequality in one variable, its solution, and the graph of the corresponding function/relation.
- Solve linear equations and inequalities numerically, graphically, and algebraically.

Investigation

Recall the problem: Two business majors, Sue and Tom, sell cookies during breaks between classes. They spend $30 on supplies and sell cookies for 25¢ each.

1. If c is the number of cookies sold and P is the net profit, write an algebraic representation (equation) for $P(c)$. What is the domain of this function?

2. Use a table of values for your equation from Investigation 1 to determine the number of cookies that must be sold to break even (to obtain a profit of $0). Record the value of c that makes the profit P zero.

3. Graph $P(c)$ using a graphing utility. Locate the horizontal intercept. Record this value.

4. Use the equation from Investigation 1 to find the horizontal intercept. Explain what you did.

5. Explain how Investigations 2–4 are related. Why is the horizontal intercept important in this problem?

6. How many cookies must be sold to make a profit? Explain.

7. How many cookies have been sold if Sue and Tom are still losing money? Explain.

Discussion

We saw that $P(c) = 0.25c - 30$ is the algebraic representation of the cookie problem. The domain of this function is limited to the whole numbers since the input is the number of cookies sold.

From Section 7.1, we know the slope is 0.25 and the vertical intercept is (0, –30). Do you remember why?

A slope of 0.25 indicates that for each one unit increase in input (each additional cookie sold), there is a $0.25 increase in the profit.

A vertical intercept of (0, –30) indicates the net profit when no cookies are sold.

The horizontal intercept is displayed numerically and graphically in Figure 1.

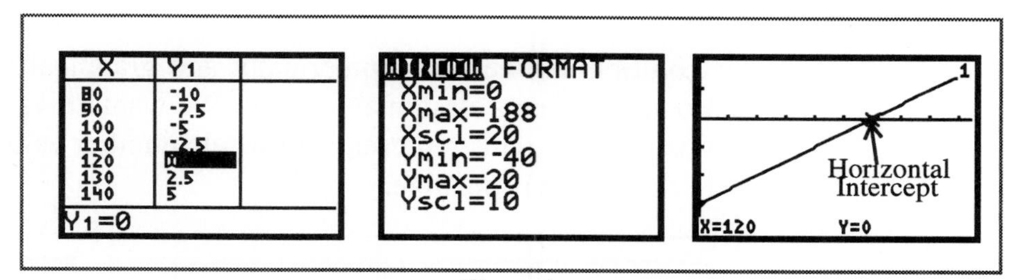

Figure 1

This is an important value since it indicates the number of cookies that must be sold if Sue and Tom hope to realize a profit of $0—this is the ***break–even point***.

In this case, we wish to find c so that the output is equal to zero. This is equivalent to the equation $P(c) = 0$. Since $P(c) = 0.25c - 30$, we get

$$0.25c - 30 = 0.$$

This is an example of an ***equation in one variable*** c. A ***solution*** to this equation is a value for c that makes the equation true.

Since the input at the horizontal intercept is 120, let's substitute 120 for c in the equation.

$$0.25\,(120) - 30 = 0$$
$$30 - 30 = 0$$
$$0 = 0$$

If we wish to know how many cookies must be sold to make a profit, we want to know the values for c where the output is greater than zero. This is equivalent to saying $P(c)$ is greater than 0. This is written algebraically as

$$P(c) > 0$$

or

$$0.25c - 30 > 0.$$

Note: The > symbol is read "greater than".

This is an example of an ***inequality in one variable*** c. A ***solution*** to this inequality is a value for c that makes the inequality true. **The** solution includes all values of c that satisfy the inequality.

Since we must sell 120 cookies to break even, we must sell more than 120 cookies to realize a profit. This is the solution to the inequality $0.25c - 30 > 0$.

This solution can be seen numerically and graphically in Figure 2. Note that no table reflects all domain values. We want all values for c where the profit is positive. This represents all whole number values on the input axis where the output is positive.

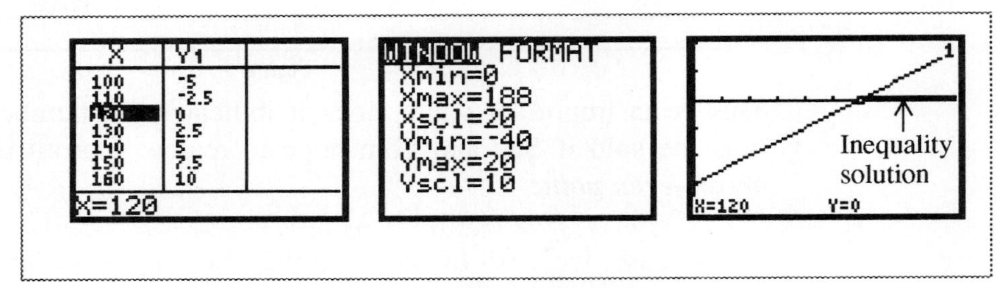

Figure 2

If we wish to know how many cookies have been sold if Sue and Tom are still losing money, we want to know the values for c where the output is less than zero. This is equivalent to saying $P(c)$ is less than 0. This is written algebraically as

$$P(c) < 0$$

or

$$0.25c - 30 < 0 \quad .$$

Note: The < symbol is read "less than".

This is another example of an ***inequality in one variable*** c. A ***solution*** to this inequality is a value for c that makes the inequality true. **The** solution includes all values of c that satisfy the inequality.

Since we must sell 120 cookies to break even, Sue and Tom lose money if less than 120 cookies are sold. This is the solution to the inequality $0.25c - 30 < 0$.

This solution can be seen numerically and graphically in Figure 3. We want all values for c where the profit is negative. This represents all whole number values on the input axis where the output is negative.

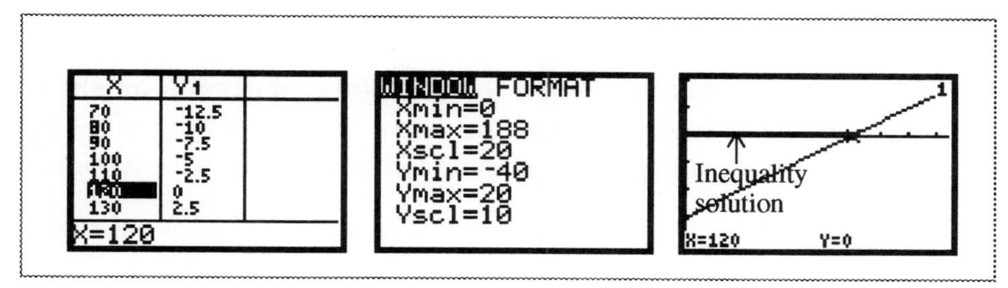

Figure 3

The equation $0.25c - 30 = 0$ and the inequalities $0.25c - 30 > 0$ and $0.25c - 30 < 0$ are examples of mathematical statements. The solutions to these statements are all values of c which make each of the statements true.

Notice that the solutions are, respectively, $c = 120$, $c > 120$, and $c < 120$. This suggests that we can determine the solution to all three of the statements by knowing the solution to any one of the statements.

In all of these situations, we've written an algebraic expression to model the problem, displayed tables, and drawn graphs. The table does not show all domain values. The algebraic representation and graphs display more values than are in the domain. That is true since these representations are assuming a domain of real numbers. The actual domain of this problem situation is limited to whole numbers.

We have found the solution to each of these statements numerically with a table and geometrically with a graph. Now we concentrate on the algebraic representation.

Investigation

8. Given the equation $0.25c - 30 = 0$, look at the left side of the equal sign, $0.25c - 30$. State the two operations being performed on the variable in the order that they would be done (computed). Be explicit.

9. Write down two operations that would undo the operations listed in Investigation 8. Indicate the order of "undoing".

10. Isolate c on the left side of the equation using the operations listed in Investigation 9. Be sure to treat both sides of the equation fairly by doing the same thing to both sides of the equation.

11. Use the same sequence of steps as in the previous investigation to isolate c in the inequality $0.25c - 30 > 0$.

12. Use the same sequence of steps as in the previous investigation to isolate c in the inequality $0.25c - 30 < 0$.

13. Solve the equation $7 - 2x = 0$ algebraically. Show and describe all steps. Check your answer.

14. How would you solve the equation $7 - 2x = 0$ using a

    a. table?

    b. graph?

15. Use your answers to Investigation 13 and 14 to write the answer to the inequality

    a. $7 - 2x < 0$.

    b. $7 - 2x > 0$.

16. Solve the inequality $7 - 2x < 0$ algebraically. Does your answer match the answer in Investigation 15?

17. a. Write a true inequality that contains only numbers.

    b. Multiply both sides your inequality by a negative number. Is your inequality still true? What must you do to make it true?

    c. What is the relationship between the result of this investigation and the steps required to solve the inequality $7 - 2x < 0$.

Discussion

Before we discuss the algebraic techniques for solving equations and inequalities, we need to summarize what we know about equations. In problems like $3 \times 4 = \quad ?$, the "=" is interpreted as a signal to do a process and get an answer. For this reason $3 \times 4 = 4 \times 3$ may cause discomfort and $5 = 5$ may not seem to be an equation because it lacks an operation. The interpretation of the "=" sign as the trigger to calculate or find an answer causes confusion with a statement like $3x + 2 = 7x - 9$.

What does this equation mean? What are the actions we are to perform? In this equation, the "=" sign is used to state the equality of two quantities. We use steps that maintain the equality, but that allow us to determine the value of the variable that makes the equation true.

In general, in *any* equation there is an "=" sign which states the equality of two quantities.

Given the equation $0.25c - 30 = 0$, the operations performed on c are

1. multiply by 0.25.
2. subtract 30.

To reverse these operations we must

1. add 30
2. divide by 0.25.

An equation is a statement that the expressions on each side of the equation are in balance. We must maintain the balance when manipulating equations algebraically. If we add a quantity to one side, we must add the same quantity to the other side. The same is true if we subtract, multiply, or divide instead of add. The same is true of inequalities. The operations we perform on one side must be performed on the other side.

To isolate c in the equation $0.25c - 30 = 0$, we must add 30 and then divide by 0.25. These operations must be performed on both sides to maintain the original balance in the equation. The process follows.

Undo the subtraction of 30 by adding 30 to **both sides** of the equation.

$$0.25c - 30 + 30 = 0 + 30$$

Simplify each side.

$$0.25c = 30$$

Undo multiplication of 0.25 by dividing **both sides** by 0.25.

$$\frac{0.25c}{0.25} = \frac{30}{0.25}$$

Simplify each side.

$$c = 120.$$

So

$$P(120) = 0.$$

The related inequalities are solved using the same steps. For example, the process to solve the inequality $0.25c - 30 > 0$ follows.

6. WeGotcha Car Rental charges \$15 per day plus 20¢ for each mile driven.

   a. If m represents the number of miles driven and T represents the total charge for one day of rental, write an algebraic representation (equation) for $T(m)$

   b. How many miles were driven if the total charge is \$40?

   c. How many miles were driven if the total charge is \$100?

7. Refer to the cookie problem at the beginning of this section.

   a. How many cookies must be sold to realize a profit of \$75?

   b. How many cookies must be sold to realize at least a \$100 profit?

8. If $y(x) = 5x + 2$, find x so that

   a. $y(x) = 14$ b. $y(x) = -4$

   c. $y(x) \le 2$ d. $y(x) > 4$

9. If $y(x) = 3 - 2x$, find x so that

   a. $y(x) = 0$ b. $y(x) = 7$

   c. $y(x) < 3$ d. $y(x) \ge 4$

10. If $y(x) = 7x + 2$, find

    a. $y(3)$ b. x if $y(x) = 3$

11. If $y(x) = 3x - 5$, find

    a. $y(-2)$ b. x if $y(x) = -2$

12. Given two linear functions $y(x) = 2x - 3$ and $y(x) = x + 4$.

    a. Write an equation that represents the situation where the outputs are equal.

    b. Solve this equation. Describe your method.

    c. Write an inequality that corresponds to the situation where the output of $y(x) = 2x - 3$ exceeds the output of $y(x) = x + 4$.

    d. Solve the inequality. Describe your method.

13. The graph of $y(x)$ is given in Figure 7. Find

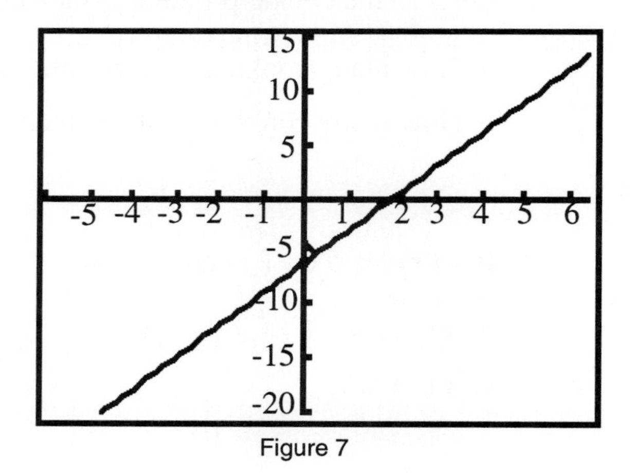

Figure 7

a. $y(1)$.

b. $y(-3)$.

c. x so that $y(x) = 0$.

d. x so that $y(x) = 9$.

e. x so that $y(x) > 5$

f. x so that $y(x) \leq -1$

Concept Map

Construct a concept map centered on the phrase **solving linear equations.**

Reflection

Compare solving a linear equation with solving a linear inequality. How are they the same? How are they different?

14. List the solutions to this equation.

Discussion

We use b and d again to represent the number of packages. The cost of b packages of burgers is $2b$. The cost of d packages of hotdogs is $1d$. If Tom spends equal amounts on burgers and hotdogs, the appropriate equation is

$$2b = d.$$

This equation has several solutions, some of which appear in Table 2. For example, the ordered pair $(4, 8)$ represents the case where Tom bought four packages of burgers and eight packages of hotdogs. We are again assuming that ordered pairs of the form (b, d) are used to represent equation solutions.

We have seen that an equation with two variables has many solutions (infinitely many if the domain is the set of real numbers), but if we have two different equations relating the two variables, we may be able to find one solution. For example, let's consider the two restrictions from the previous investigations simultaneously. Tom must spend the entire $20 and he must spend equal amounts on burgers and hotdogs. This requires that we find any common solutions to the two equations

$$\begin{aligned} 2b + d &= 20 \\ 2b &= d \end{aligned}.$$

We will look at numerical, graphical, and algebraic methods of solutions. When several equations relate the same variables, we call them a ***system of equations***. Because we are looking for a solution that satisfies all the equations at the same time, they are also called ***simultaneous equations***.

Each equation expresses a functional relationship between the variables b and d. Since we used the ordered pair (b, d), we are assuming that the number of packages of burgers b is the input and the number of packages of hotdogs d is the output. To analyze the system numerically and graphically, we must first express d as a function of b in each equation. Let's do so.

Investigation

15. The equation $2b = d$ already expresses d as a linear function of b. Identify the slope and vertical intercept for this function.

16. The equation $2b + d = 20$ must be solved for d.

    a. What operation is being performed on d on the left side of the equation?

    b. What must be done to "undo" this operation?

    c. Use equation solving techniques to isolate d on one side.

    d. The result is a linear function. Identify its slope and vertical intercept.

17. Enter both functions on your calculator. Display a table of outputs for both functions. What is the common solution to the two equations? How do you know?

18. Graph both equations on your calculator in a "friendly" window. Determine where the graphs intersect.

19. Explain why the point of intersection is the same ordered pair solution found in the table.

20. How is this problem similar to the monk's journey problem? How is it different?

21. We will use a technique called substitution to solve the problem algebraically.

    a. Use the fact that $d = 2b$ to substitute for d in the equation $2b + d = 20$. Write the resulting equation.

    b. The equation from part a has only one variable *b*. Use equation solving techniques to solve for *b*.

    c. Substitute your answer from part b into either of the original equations. Solve for *d*.

    d. Do your answers for *b* and *d* match those obtained using a table and a graph? If not, check your work.

Discussion

The equation $d = 2b$ has a slope of 2 and a vertical intercept at (0, 0).

To solve the equation $2b + d = 20$ for *d*, we note that $2b$ is being added to *d*. We "undo" this step by subtracting $2b$ from **both sides**.

$$2b + d - 2b = 20 - 2b$$

Simplify and rewrite in a more standard form.

$$d = 20 - 2b$$

$$d = -2b + 20$$

This function has a slope of –2 and a vertical intercept at (0, 20).

The solution to the system of equations is found using the table and the graph in Figure 2.

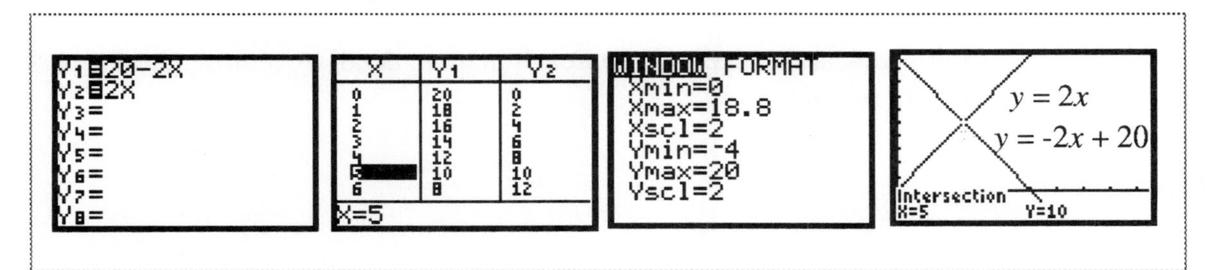

Figure 2

Note that *d* and *b* have been replaced with *x* and *y* on the calculator.

The algebraic method that we will use to solve this system is called ***substitution***. Substitution is an important principal in mathematics that states that if two quantities are equal, one may be substituted for the other in any expression.

Consider the two equations we have been using.

$$2b + d = 20 \text{ and } d = 2b$$

Since $d = 2b$ in the second equation, we will replace d with $2b$ in the first equation. This gives

$$2b + 2b = 20.$$

Since this is a linear equation, there is one solution for b.

$$4b = 20$$

$$b = 5$$

We still must find the value for d. Since $d = 2b$ and $b = 5$,

$$d = 2(5) = 10.$$

So Tom should buy 5 packages of burgers and 10 packages of hotdogs if he wishes to spend an equal amount of money on burgers and dogs.

Summarizing, we solved this system of equations

- numerically by displaying a table of outputs for both functions. The place where the outputs were equal indicated the solution to the system.
- graphically by displaying the graphs of both functions and finding the intersection point.
- algebraically using a technique called substitution.

Let's look at the algebraic approach in more detail by returning to the parking lot problem.

Investigation

Sometimes substitution is not so obvious. In Section 1.1, we indicated that the parking lot problem could be solved using a system of equations. The system we suggested was

$$c + m = 20 \text{ and } 4c + 2m = 66.$$

The c represents the number of cars parked in the lot and the m represents the number of motorcycles parked in the lot. The first equation expresses the fact that the total number of vehicles in the lot is 20. The second equation expresses the fact that the total number of wheels in the lot is 66.

First you must choose one equation and one variable that you will solve for. Theoretically we have four choices in this system of equations. Practically we want to make a choice that makes the algebraic manipulations as easy as possible.

22. Which equation and variable would you choose? Why?

23. Solve the equation you chose for the variable you selected.

24. Substitute your answer from Investigation 23 into the other equation and solve for the remaining variable.

25. Find the ordered pair that satisfies both equations. Be sure to check your answer using both a table and a graph.

Discussion

To simplify manipulations, we choose the variable with a numerical coefficient of one. It is c or m in the equation $c + m = 20$.

We choose to solve this equation for c.

Subtract m from both sides to get $c = 20 - m$.

Substitute the expression $(20 - m)$ for c into the equation

$$4c + 2m = 66$$

$$4(20 - m) + 2m = 66$$

Note that $(20 - m)$ is in parentheses so that 4 distributes over the quantity.

Distribute the 4 to remove parentheses: $80 - 4m + 2m = 66$

Combine like terms: $80 - 2m = 66$

Subtract 80 from both sides: $-2m = -14$

Divide both sides by –2: $m = 7$

Substitute $m = 7$ into $c = 20 - m$: $c = 20 - (7) = 13$

So the pair $m = 7$ and $c = 13$ should satisfy both equations.

Let's check our solution by substituting the pair into both equations to see if true statements result.

Substitute $m = 7$ and $c = 13$ into $c + m = 20$

$$13 + 7 = 20$$

$$20 = 20$$

Substitute $m = 7$ and $c = 13$ into $4c + 2m = 66$

$$4(13) + 2(7) = 66$$

$$42 + 14 = 66$$

$$66 = 66$$

Since the substitution of $m = 7$ and $c = 13$ results in a true statement for both equations, we know the pair is the solution to the system.

There is another common technique, often called elimination, that is used to solve linear systems. That is discussed in another text.

In Section 7.3, we solved linear equations and inequalities where the variable appeared on only one side of the equation. Let's use systems to look at how to solve linear equations and inequalities where the variable appears on both sides.

Investigation

Consider the linear functions $y = 2x + 3$ and $y = 5x - 9$.

26. Use a table to find the input where the outputs are equal. Write the answer along with the ordered pair that satisfies both equations.

27. Graph both functions. What point represents the common solution to both equations?

28. Use substitution to write one equation that represents the situation where the outputs are equal.

Discussion

If we wish to know the input where the outputs of $y = 2x + 3$ and $y = 5x - 9$ are the same, we are trying to find the solution to the system of equations

$$\begin{aligned} y &= 2x + 3 \\ y &= 5x - 9 \end{aligned}.$$

Figure 3 displays both a numerical and graphical solution.

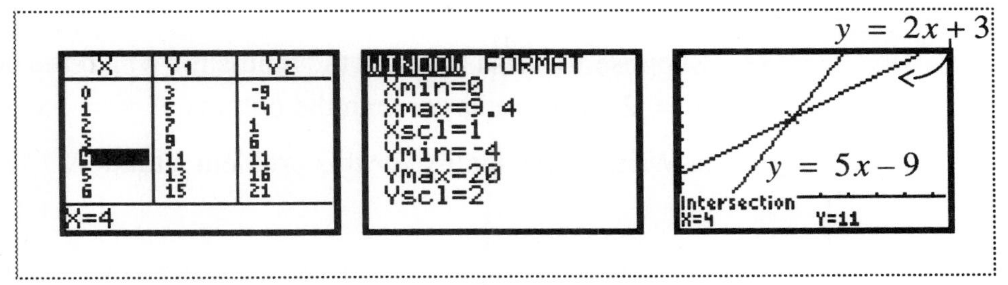

Figure 3

From both the table and the graph, we see that the solution to the system is the ordered pair $(4, 11)$.

We can use substitution by setting the processes of the two functions equal. Thus we have an equation in x only.

$$2x + 3 = 5x - 9$$

Let's investigate how to solve this equation algebraically. Remember that we must use steps that maintain the balance of the equation.

Investigation

29. We would like to combine the two terms containing x into one term. To do so, both terms must be on the same side of the equal relation. What can be done to both sides so that terms with x appear on one side only? Perform this operation and write the resulting equation.

30. Solve the equation from the previous investigation

31. Suppose we wish to know the inputs for which the outputs of $y = 2x + 3$ are smaller than the outputs of $y = 5x - 9$.

    a. Write an inequality for this problem situation.

    b. Write a conjecture about the answer to this inequality by using the answer to the equation along with Figure 3.

    c. Solve the inequality algebraically. Check your answer using either a table or a graph.

32. Suppose we wish to know the inputs for which the outputs of $y = 2x + 3$ are larger than the outputs of $y = 5x - 9$.

    a. Write an inequality for this problem situation.

    b. Write a conjecture about the answer to this inequality by using the answer to the equation along with Figure 3.

    c. Solve the inequality algebraically. Check your answer using either a table or a graph.

Discussion

If terms containing the variable appear on both sides of an equation, we subtract one of the terms from both sides of the equation. This leaves us with an equation with a variable on only one side.

Original equation	$2x + 3 = 5x - 9$
Subtract $5x$ from **both sides**	$2x + 3 - 5x = 5x - 9 - 5x$

Note that we could have chosen to subtract $2x$ from both sides. This results in the same solution.

Simplify each side $\quad -3x+3 = -9$

Now we must isolate x by "undoing" what was done to the variable.

Subtract 3 from **both sides** $\quad -3x+3-3 = -9-3$

Simplify each side $\quad -3x = -12$

Divide **both sides** by –3 $\quad \dfrac{-3x}{-3} = \dfrac{-12}{-3}$

Simplify each side $\quad x = 4$

If we substitute 4 for x in either equation, we get an output of 11. The solution to the system is $(4, 11)$.

The inequality given by the condition that the output of $y = 2x+3$ is to be smaller than the output of $y = 5x-9$ is

$$2x+3<5x-9.$$

From our previous solutions and Figure 3, we conjecture that the solution is $x>4$. Let's solve algebraically for the practice.

Original inequality $\quad 2x+3<5x-9$

Subtract $5x$ from **both sides** $\quad 2x+3-5x<5x-9-5x$

Simplify each side $\quad -3x+3<-9$

Subtract 3 from **both sides** $\quad -3x+3-3<-9-3$

Simplify each side $\quad -3x<-12$

Divide **both sides** by –3. Remember that the direction of the inequality must reverse since we are dividing by a negative number.

$$\frac{-3x}{-3}>\frac{-12}{-3}$$

Simplify each side $\quad x>4$

The inequality given by the condition that the output of $y = 2x + 3$ to be larger than the output of $y = 5x - 9$ is

$$2x + 3 > 5x - 9.$$

This can be solved in the same manner.

Figure 4 displays graphical solutions to both inequalities.

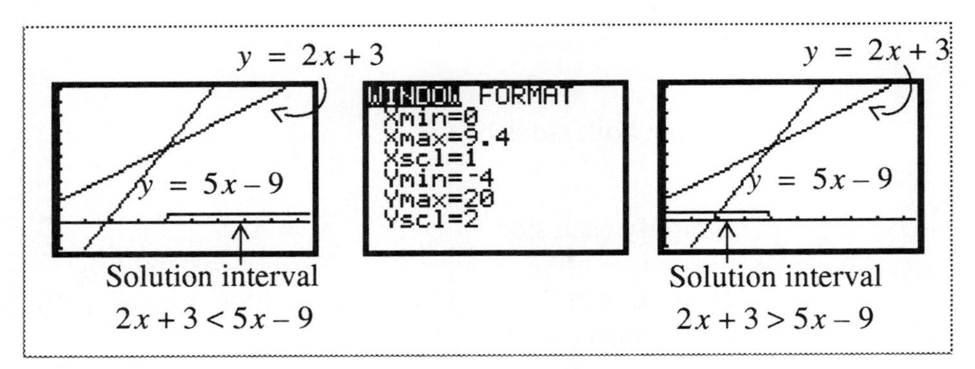

Figure 4

Explorations

1. List and define the words in this section that appear in ***italics bold*** type.

2. Find the solutions graphically, numerically, using a table, and by substitution.

   a. $x + y = 8$, $x - y = 4$

   b. $x + 2y = 5$, $x + y = 3$

   c. $5x - y = 13$, $2x + 3y = 12$

   d. $x + 3y = 5$, $2x - 3y = -8$

   e. $3x - 2y = 0$, $5x + 10y = 4$

   f. $2x + 3y = 5$, $4x + 6y = 10$

   g. $2x + 3y = 4$, $6x + 9y = 18$

3. The equations in Exploration 2f are called ***dependent equations***. Why? What happened when you tried to solve the system of equations?

4. The equations in Exploration 2g are called ***inconsistent equations***. Why? What happened when you tried to solve the system of equations? How is this problem different from the system of equations in Exploration 2f?

5. Write a story problem involving two variables with two relationships (equations) that you experienced this week. Identify the variables and write appropriate mathematical statements that you can use to solve the problem.

6. I went to the store and bought the same number of books as cassettes. Books cost two dollars each and cassettes cost six dollars each. I spent $40 all together. Assuming that the equation $2B + 6R = 40$ is correct, what is wrong, if anything, with the following reasoning? Be as detailed as possible.

$$2B + 6R = 40$$

Since $B = R$ in this problem, I can write

$$2B + 6B = 40 \text{ Why?}$$
$$8B = 40$$
$$B = 5$$

Since the equation $8B = 40$ says 8 books cost $40, one book costs $5.

7. Find the solution to each of the following equations or inequalities.

   a. $2x - 3 = x + 4$
   b. $7x + 2 = 4x - 5$
   c. $x - 8 > 2 - 3x$
   d. $5x + 4 \leq 3x$

8. In Exploration 7 in Section 1.1, you found several ways to solve the basketball problem. Use the techniques of this section to solve numerically, algebraically, and graphically.

9. Consider the linear functions $y = 3x - 7$ and $y = 4 - 5x$.

   a. Assume the outputs are equal. Write an equation that represents this fact. Solve the equation.

   b. Assume that the output of $y = 3x - 7$ is no larger than the output of $y = 4 - 5x$. Write an inequality that represents this fact. Solve the inequality.

Concept Map

Construct a concept map centered on the phrase **systems of equations**.

Reflection

Based on your concept map, write a paragraph describing what you know about systems of equations.

Section 7.5 Finding Zeros of Quadratic Functions by Factoring

Purpose

- Investigate factoring involving the greatest common factor.
- Investigate factoring trinomials with a leading coefficient of one.
- Use tables and graphs as an aid in factoring.

Investigation

Consider the problem: We wish to build a rectangular dog pen in which the length is 3 yards more than the width.

1. If x yards represents the width of the pen, write an algebraic expression for the length of the pen.

2. Complete Table 1.

Width	Length	Area
2		
5		
10		
12		
17		
x		

Table 1: Building a Dog Pen

3. If x represents the width and A represents the area, write an algebraic representation (equation) for $A(x)$.

4. Graph the function from Investigation 3 on a graphing utility. Determine the width so that the area is 18 square yards.

5. Use a table to verify your answer(s) to Investigation 4. What do you look for in the table to solve this problem?

Discussion

The area of the dog pen is represented by the equation $A(x) = x(x+3)$ in factored form or $A(x) = x^2 + 3x$ in expanded form.

The table and graph of the function appears as Figure 1.

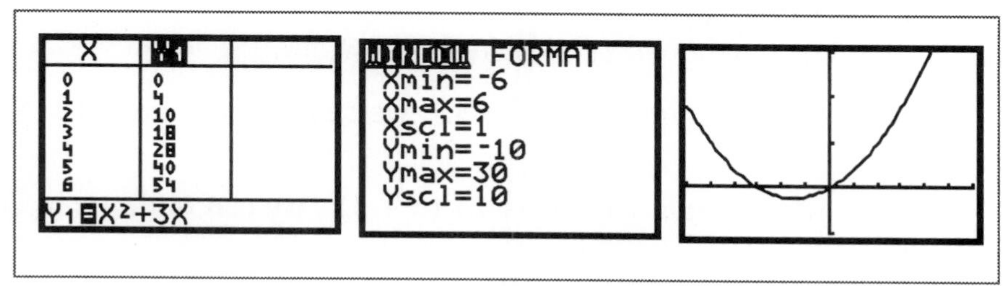

Figure 1

In the table, identify the input where the output is 18.

On the graph, the horizontal line through an output of 18 intersects the graph at two input values. One of these values—the positive value—tells us the width required to obtain an area of 18 square yards.

In Investigation 4, we looked at the problem graphically. In Investigation 5, we studied the problem numerically. Now let's look at the problem algebraically. We wish to know when $A(x) = 18$. This is equivalent to the equation

$$x^2 + 3x = 18.$$

Subtracting 18 from both sides to make one side zero, we get

$$x^2 + 3x - 18 = 0.$$

We need a method of solving this equation. We could do it numerically by guessing values for x that will make this equation true. This is equivalent to entering the function $y(x) = x^2 + 3x - 18$ in the calculator and looking in the table for an output of zero. Give it a try!

We could solve it graphically by graphing $y(x) = x^2 + 3x - 18$ and finding the horizontal intercepts.

To solve algebraically, we need to look at factoring the quadratic polynomial.

Factoring reverses the process of multiplying. For example, if we write $(2)(3) = 6$, we multiplied the ***factors*** of 2 and 3 to obtain a **product** of 6. If we reverse the process, $6 = (2)(3)$, we ask ourselves what two numbers have a product of 6 and express 6 as a product of these factors.

We can look at polynomials in the same context.

For example, if we begin with $x(2x+3)$, we can multiply to get $2x^2 + 3x$.

If we begin with $2x^2 + 3x$, we can factor to get $x(2x+3)$

If we begin with $(x+2)(x+3)$, we can multiply to get $x^2 + 5x + 6$.

If we begin with $x^2 + 5x + 6$, we can factor to get $(x+2)(x+3)$.

A factorization in one variable can be checked by entering the original polynomial as one function and the factored form as a second function. The table outputs should match if the factored form is equivalent to the original polynomial. Figure 2 displays a table that uses $x^2 + 5x + 6$ for Y_1 and $(x+2)(x+3)$ for Y_2.

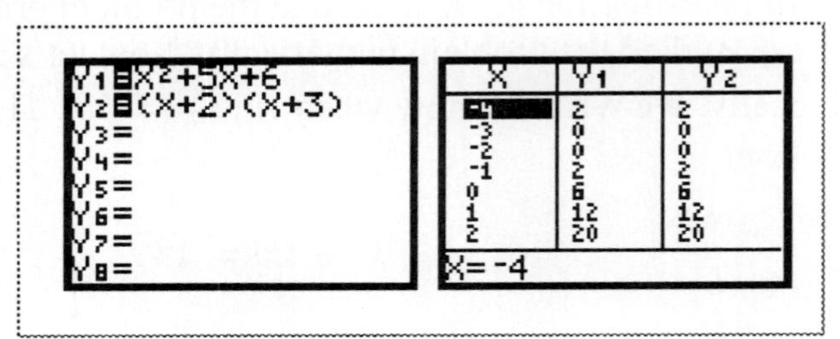

Figure 2

If a polynomial does not factor, it is called a ***prime polynomial***.

In Section 7.2, we explored the relationships between the zeros of a quadratic function and the factors of a quadratic function. Knowing the zeros suggests the factorization. For example, the data in Table 2 shows that the function $y(x) = x^2 + 5x + 6$ has zeros of –3 and –2. These are the inputs where the output is zero.

Factors have the general form $(x - (\text{zero of polynomial}))$.

So if –3 and –2 are zeros of $x^2 + 5x + 6$, then $(x - (-3)) = (x + 3)$ and $(x - (-2)) = (x + 2)$ are factors of $x^2 + 5x + 6$.

We now will explore some algebraic techniques for factoring quadratic functions. The first technique we consider involves using the distributive property of multiplication over addition or subtraction to rewrite polynomials of the form $ab + ac$ as $a(b + c)$. This technique is called ***removing the common monomial factor***.

Investigation

6. Predict how each of the following will factor. Use a symbol manipulator to factor and compare the answers to your predictions.

Expression	Prediction	Symbol Manipulator
$x^2 + 5x$		
$x^2 - 9x$		
$3a + 6$		
$18b - 12$		
$6c^2 + 4c$		
$12x^2 - 4x$		

Table 2: Factoring 1

7. Factor by hand. Check your answers using both a symbol manipulator and a table.

   a. $x^2 + 5x$

   b. $5y - 15$

   c. $8b^2 - 14b$

   d. $10x^2 + 5x$

8. Write a statement explaining how you factored the problems in Investigation 7.

Discussion

If every term of a polynomial contains a common factor, we can use the distributive property of multiplication over addition or subtraction to factor the polynomial. We rewrite the polynomial as the product of the common factor and the sum of the remaining factors. Let's look at the ones that you did.

$$x^2 + 5x = x(x) + x(5) = x(x+5)$$

Each term has a common factor of x.

$$5y - 15 = 5(y) - 5(3) = 5(y-3)$$

Each term has a common factor of 5.

$$8b^2 - 14b = 2b(4b) - 2b(7) = 2b(4b-7)$$

Each term has a common factor of $2b$.

$$10x^2 + 5x = 5x(2x) + 5x(1) = 5x(2x+1)$$

Each term has a common factor of $5x$.

Figure 3 displays a check of the last factorization using a table.

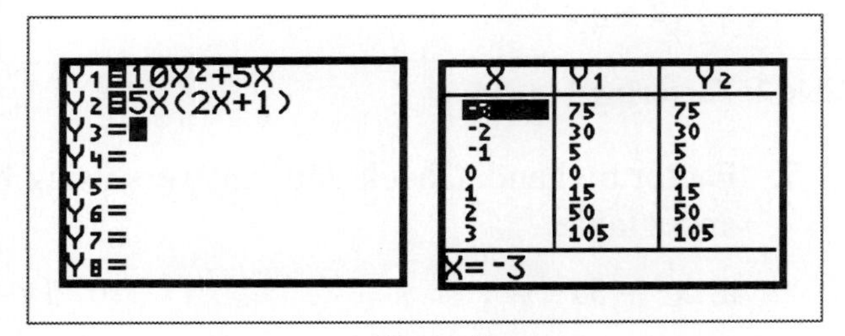

Figure 3

Note that the outputs for the original polynomial and the factored form are the same. This suggests that our factorization is correct.

We now will investigate the factorization of quadratic trinomials. We will use a table to suggest the zeros. We will create the factors from the zeros. Finally, we will check the factorization using a symbol manipulator.

Investigation

9. Use a table or symbol manipulator to identify the zeros of each function. Use the zeros to predict the factorization. Use a symbol manipulator to factor and compare the answers to your predictions.

Expression	Zeros	Predicted factorization	Factorization on symbol manipulator
$x^2 + 5x + 6$			
$x^2 + 11x + 28$			
$x^2 + 16x + 55$			
$x^2 + 17x + 72$			
$x^2 + 10x + 24$			
$x^2 + 22x + 117$			

Table 3: Factoring 2

10. Each expression in Table 3 has the form $x^2 + px + q$ where p and q are positive numbers.

    a. What are the signs of the zeros?

    b. How do the zeros relate to the constant term?

    c. How do the zeros relate to the numerical coefficient of x?

    d. Generalize the form of the factorization.

11. Use a table or symbol manipulator to identify the zeros of each function. Use the zeros to predict the factorization. Use a symbol manipulator to factor and compare the answers to your predictions.

Expression	Zeros	Predicted factorization	Factorization on symbol manipulator
$x^2 - 5x + 6$			
$x^2 - 11x + 28$			
$x^2 - 16x + 55$			
$x^2 - 17x + 72$			
$x^2 - 10x + 24$			
$x^2 - 22x + 117$			

Table 4: Factoring 3

12. Each expression in Table 4 has the form $x^2 - px + q$ where p and q are positive numbers.

    a. What are the signs of the zeros?

    b. How do the zeros relate to the constant term?

    c. How do the zeros relate to the numerical coefficient of x?

    d. Generalize the form of the factorization.

13. Factor by hand. Check your answers using both a symbol manipulator and a table.

    a. $x^2 + 7x + 12$

    b. $x^2 + 3x + 2$

    c. $x^2 - 13x + 36$

    d. $x^2 - 23x + 132$

    e. $x^2 - 15x + 36$

Discussion

We wish to factor a quadratic of the form $x^2 + px + q$ where p and q are positive integers. Here are some generalizations we can draw from Table 3.

- Both zeros will be negative, if they are real numbers.
- The product of the two zeros is the constant term.
- The absolute value of the sum of the two zeros is the coefficient of x.
- The factorization will be of the form $(x + a)\ (x + b)$ where $-a$ and $-b$ are the zeros.

If the quadratic has the form $x^2 - px + q$ where p and q are positive integers, some generalizations from Table 4 follow.

- Both zeros will be positive, if they are real numbers.
- The product of the two zeros is the constant term.
- The sum of the two zeros is the absolute value of the coefficient of x.
- The factorization will be of the form $(x - a)\ (x - b)$ where a and b are the zeros.

Let's look at the problems you did.

$x^2 + 7x + 12$

Both zeros are negative.

The table and graph appear in Figure 4.

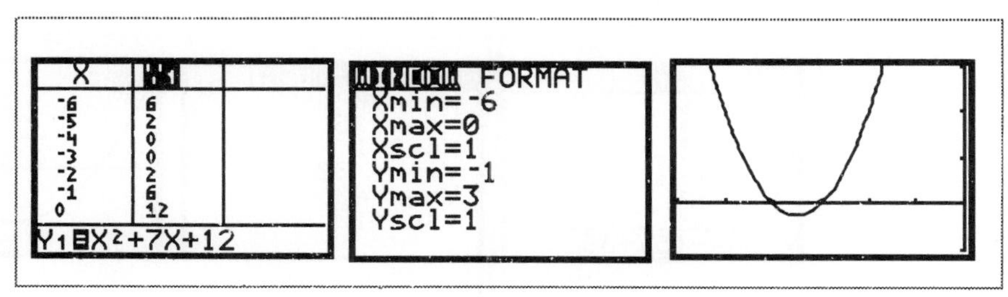

Figure 4

The zeros of this function are –3 and –4.

The factors are $(x - (-3))$ and $(x - (-4))$.

Therefore $x^2 + 7x + 12 = (x + 3)\ (x + 4)$.

$x^2 + 3x + 2$

Both zeros are negative.

Using a table or graph, we note that the zeros are –1 and –2.

In this problem, $p = 3$ and $q = 2$. The product of the zeros is 2 and the absolute value of their sum is 3.

If –1 and –2 are zeros, then $(x - (-1))$ and $(x - (-2))$ are factors.

Therefore $x^2 + 3x + 2 = (x + 1)(x + 2)$.

$x^2 - 13x + 36$

Both zeros are positive.

The zeros are 4 and 9.

The factors are $(x - 4)$ and $(x - 9)$.

Therefore $x^2 - 13x + 36 = (x - 4)(x - 9)$.

$x^2 - 23x + 132$

Both zeros are positive.

The zeros are 11 and 12.

The factors are $(x - 11)$ and $(x - 12)$.

Therefore $x^2 - 23x + 132 = (x - 11)(x - 12)$.

$x^2 - 15x + 36$

Both zeros are positive.

The table and graph appear in Figure 5.

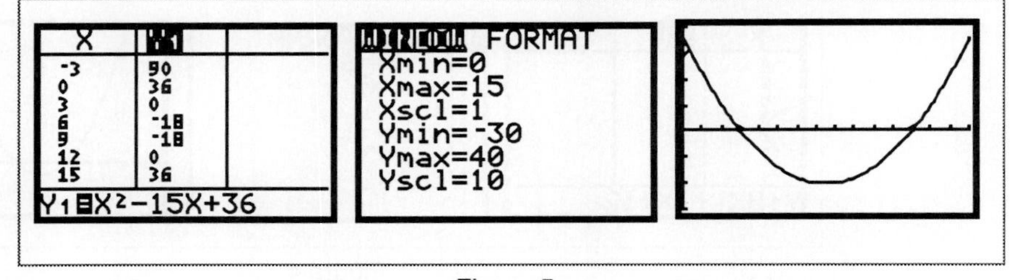

Figure 5

The zeros of this function are 3 and 12.

The factors are $(x - 3)$ and $(x - 12)$.

Therefore $x^2 - 15x + 36 = (x - 3)(x - 12)$.

The zeros are –6 and 3. Since these values represent the width of the rectangle, we consider only the positive answer. Therefore, to obtain an area of 18 square yards, the width of the pen must be 3 yards and the length of the pen must be 6 yards.

Explorations

1. List and define the words in this section that appear in ***italics bold*** type.

2. Find the zeros and write the factorization, if possible. If the quadratic function does not factor over the rational numbers, write prime.

Function	Zeros	Factorization
$x^2 - 25$		
$x^2 - 16$		
$x^2 + 25$		
$x^2 + 16$		
$x^2 - 5$		
$x^2 + 5$		
$4x^2 - 9$		
$4x^2 + 9$		
$2x^2 - 9$		

Table 6: Quadratic binomials

3. Assume c is a non–zero whole number. Study the results of Table 6.

   a. Under what condition on c will $x^2 - c$ factor over the rational numbers?

   b. Does $x^2 + c$ factor over the rational numbers? Why or why not?

4. Consider the quadratic polynomials $x^2 - 12x + 36$, $x^2 + 14x + 49$, $x^2 - 6x + 9$, and $x^2 + 8x + 16$.

   a. Find the zeros and factor each.

   b. How many unique zeros does each quadratic have?

   c. How many different factors does each quadratic have?

   d. What do you notice about how the graphs of the functions intersect the x–axis?

e. What do you notice about the first and last terms of these quadratics?

f. How is the middle term related to the first and last terms?

5. The graph of a quadratic function appears in Figure 10.

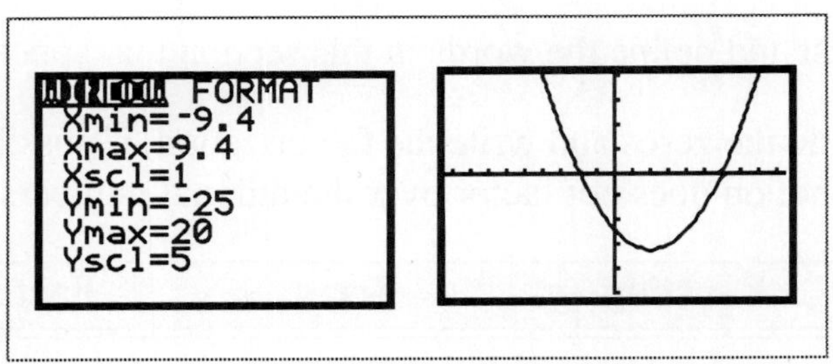

Figure 10

a. What are the zeros of this function?

b. What are the factors of this function?

c. Write the algebraic representation of this function.

6. The graph of a quadratic function appears in Figure 11.

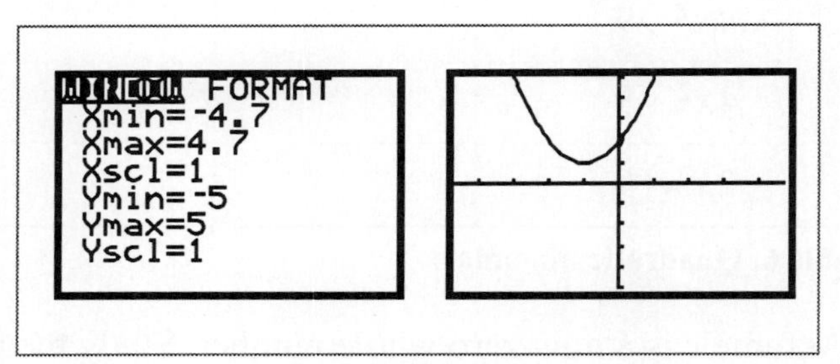

Figure 11

a. What are the zeros of this function?

b. Does this function factor over the rational numbers? Why or why not?

7. Here's some practice, if you need it. Find the zeros and factor completely. If prime, say so.

a. $9x^2 - 18x$

b. $21y^4 - 14y^3$

c. $5x^3 + 25x^2 - 30x$

d. $x^2 + 9x + 8$

e. $p^2 - 6p - 7$

f. $x^2 + 12x + 13$

g. $x^2 - x - 30$

h. $y^2 - 2y - 45$

i. $x^2 + 8x + 15$

j. $x^2 - 6x + 8$

k. $x^2 + 9x - 22$

l. $p^2 + 2p - 15$

m. $x^2 - x - 56$

n. $x^2 - 15x + 44$

o. $x^2 + 14x + 49$

p. $y^2 - 12y + 36$

q. $x^2 + 12x - 35$

r. $2y^2 - 10y - 48$

s. $24x + 20$

t. $t^2 - 17t + 66$

u. $x^2 + 17x + 72$

v. $16x^2 + 48x + 36$

8. For the dog pen area function $A(x) = x^2 + 3x$ studied earlier, determine the dimensions if the area is:

   a. 70 square yards.

   b. 108 square yards.

   c. 4 square yards.

9. A rectangle has a length that is 7 feet more than its width.

   a. If x feet represents the width, write an expression for the length.

   b. If x feet represents the width and A square feet represents the area, write an algebraic representation (equation) for $A(x)$.

   c. Find the area of the rectangle if the width is 9 feet.

   d. Find the area of the rectangle if the length is 12 feet.

   e. Find the width and length of the rectangle if the area is 78 square feet.

Concept Map

Construct a concept map centered on the word **factoring**.

Reflection

Write a paragraph describing what you know about factoring.

Section 7.6 Making Connections: Linking Multiple Representations of Functions

Purpose

- Reflect upon ideas explored in Chapter 7.
- Explore the connections among multiple representations of functions.

Investigation

In this section you will work outside the system to reflect upon the mathematics in Chapter 7: what you've done and how you've done it.

1. State the five most important ideas in this chapter. Why did you select each?

2. Identify all the mathematical concepts, processes and skills you used to investigate the problems in Chapter 7?

3. What has been common to all of the investigations which you have completed?

4. Select a key idea from this chapter. Write a paragraph explaining it to a confused best friend.

5. You have investigated many problems in this chapter.

   a. List your three favorite problems and tell why you selected each of them.

   b. Which problem did you think was the most difficult and why?

6. Explain what it means to "find a solution" numerically, algebraically and graphically.

Discussion

There are a number of really important ideas which you might have listed including equation, root, solution, graphic solution, systems of equations, dependent equations, and inconsistent equations.

Concept Map

Construct a concept map centered around the phrase **doing mathematics**.

Reflection

Before beginning this class, what was your idea of "doing mathematics"? How has your work in this course altered this idea?

Illustration

Draw a picture of a student evolving into a mathematician.

The Student
Mathematician

Glossary

A

absolute value function

addends

additive identity

additive inverse

algebraic order of operations

algebraic representation

associative property of addition

associative property of multiplication

axis

B

base

binary

binary operation

binomial

break–even point

C

combining like terms

commutative property of addition

commutative property of multiplication

constant

cumulative sums

D

degree

delta

denominator

dependent variable

difference of two squares

direct variation

discrete

distributive property of multiplication over addition

distributive property of multiplication over subtraction

divisor

domain

domain of a function.

E

equation

equation in one variable

equivalent

evaluate

exponential expression

F

factors

finite differences

function

function machine

function notation

G

general linear function

general quadratic function

graphing utility

H

horizontal intercept

I

identity function

identity property for multiplication

independent variable

index

infinity

initial value

input

integers

inverse variation

is equivalent to

L

like terms

linear function

linear inequality

linear polynomial

M

magnitude

monomial

multiplicative identity

multiplicative inverse

N

nonterminating, nonrepeating decimal

numerator

numerical coefficient

O

opposite

oppositing

ordered pair

origin

output

P

parabola

parametric equations

piecewise–defined function

plane

point

polynomial term

polynomials

prime polynomial

principal square root

Q

quadratic polynomial

R

radical

radicand

range

range of the function

rate of change

rational numbers

real number system

reciprocal

relation

S

sequence

simultaneous equations

slope

solution

square trinomial

subscripts

substitution

subtrahend

symbol of inclusion

symmetry

system of equations

T

term

terminating

triangular numbers

trinomial

truncates

U

unary

unary operation

unary operator

V

variables

vary directly

vary inversely

vertex

vertical intercept

viewing rectangle

viewing window

X

x–intercept

Xmax

Xmin

Xscl

Y

y–intercept

Ymax

Ymin

Yscl

Z

zeros of the function

Index

Index

Index